- Why is addressing the sources contentious
  - arguments
- Who should pay

CLIMATE CHANGE LAW

SECOND EDITION

DANIEL A. FARBER
Sho Sato Professor of Law
University of California, Berkeley

CINNAMON P. CARLARNE
Associate Dean for Faculty & Intellectual Life and
Robert J. Lynn Chair in Law
Michael E. Moritz College of Law,
The Ohio State University

CONCEPTS AND INSIGHTS SERIES®

860 Blue Gentian Road, Suite 350
Eagan, MN 55121
1-877-888-1330
Printed in the United States of America

ISBN: 978-1-63659-626-6

PREFACE

Over the past thirty years, a body of law dealing with the challenge of climate change has taken form. This rapidly emerging body of law runs the gamut from state and local regulations to federal policies and international agreements and includes both public and private sector involvement. It implicates questions of mitigation and adaptation, including complex questions at the intersection of law and politics, science, economics, technology, and equity.

Our goal in writing this book is to provide a concise but careful and timely overview of this important and quickly changing area of law. The book is organized as follows. It begins by discussing the scientific and policy issues that frame the legal scheme, including the state of climate science, the meaning of the social cost of carbon, and the variety of tools that are available to reduce carbon emissions. It then covers in turn the international, national, and state efforts in this sphere. Finally, the book turns to the challenge of adapting to climate change and the controversial possibility of using geoengineering as a tool for addressing climate change.

Our strategy throughout these chapters is to offer readers the information and tools they need to understand climate change law in its present form, and to keep pace as it continues to evolve. This is particularly important because, as this book goes to press, we are the precipice of what is sure to be an eventful few years in the development of climate law. President Biden recently signed a historic federal climate bill, the 2022 Inflation Reduction Act (IRA). The IRA provides for unprecedented investments in climate action, notably $369 billion in tax credits and direct spending to support clean energy and climate resilience. It is estimated that the IRA will help the United States reduce domestic carbon emissions by roughly 40% by 2030. As a result, the existing body of climate law is likely to grow at an accelerated pace over the coming years. That growth is sure to pose legal issues, as demonstrated by a Supreme Court's decision (*West Virginia v. EPA*) issued just a couple months before the IRA that limited EPA's authority to cut emissions.

We hope that this book is useful and accessible to a broad range of readers, including those who do not have backgrounds in climate science, environmental economics, or law. We are grateful to have the opportunity to engage with readers on this important issue and we hope that you find this book helpful and informative!

We owe thanks to users of the first edition for their helpful feedback. We also wish to thank Jetta Cook (Berkeley JD 2022), Erin Ahern (OSU JD class of 2024), Kaylin Shackelford (OSU JD class of 2024), and Gwen Short (OSU JD Class of 2022) for their research assistance.

Dan Farber
Cinnamon Carlarne

08.15.2022

ACKNOWLEDGMENTS

C.P.C. thanks Spencer Kluth for her assistance, and dedicates her portion of the book to Keith Hirokawa and Cobu who keep me inspired and truckin' along, and to my daughter, Matilda Carlarne, who is my joy and my motivation for fighting for a safer and more equitable future.

D.A.F. thanks Jetta Cook for her assistance, and dedicates his portion of the book to Emma Farber, Helene Farber Gillespie, Henry Farber Gillespie, and June Farber Allen, who are representatives of those who will be most affected by climate change in coming decades.

SUMMARY OF CONTENTS

TABLE OF CONTENTS

TIME LINE

1983 Iowa is the first state to establish a renewable portfolio standard (RPS).

1988 U.N. establishes the Intergovernmental Panel on Climate Change (IPCC).

1990 First IPCC report, finding that human activities contribute to global warming.

1992 UNFCC signed.

1994 UNFCC enters into force.

1995 Second IPCC report, confirming a discernible human influence on global climate.

1997 Kyoto Agreement signed.

2001 Third IPCC report, finding that most of the warming of the past 50 years is likely to be attributable to human activities.

2002 California passes Pavley Act regulating carbon emissions from new cars.

2005 Kyoto Agreement enters into force without U.S.

Northeastern states establish Regional Greenhouse Gas Initiative (RGGI).

2006 Supreme Court decides *Massachusetts v. EPA.*

California adopts AB32 climate law.

EPA approves California mandate for zero emission vehicles.

2007 Fourth IPCC report, finding that warming is unequivocal, and that most of the warming of the past 50 years is very likely due to human induced increases in greenhouse gases.

2009 EPA formally finds that greenhouse gas emissions endanger human health and welfare.

House of Representatives passes Waxman-Markey climate bill, which dies in the Senate.

Copenhagen Accord agreed.

2011 Supreme Court decides *AEP* case.

2012 EPA adopts standard for carbon emissions from new cars.

2014 Supreme Court decides *UARG v. EPA.*

Fifth IPCC report, finding that human influence on the climate system is clear, that recent anthropogenic emissions of greenhouse gases are the highest in history, and that recent climate changes have had widespread impacts on human and natural systems.

2015 Paris Agreement adopted.

Clean Power Plan issued, then stayed by Supreme Court.

2016 Paris Agreement enters into force.

2017 Trump announces U.S. withdrawal from Paris Agreement.

2018 Hawaii becomes first state to adopt a binding net-zero target.

2019 EPA issues Affordable Clean Energy rule, repeals Clean Power Plan.

2020 U.S. withdraws from Paris Agreement (Trump).

2021 U.S. rejoins Paris Agreement (Biden).

D.C. Circuit vacates Affordable Clean Energy rule.

Glasgow Climate Pact Adopted.

2022 Sixth IPCC report, finding that it is "unequivocal" that human influence has warmed the atmosphere, oceans, and land; that the path to limiting warming to 1.5C is increasingly narrow; and that climate change is a threat to human health, global equity, and economic development.

EPA issues new standards for carbon emissions from new vehicles.

Supreme Court decides *West Virginia v. EPA*, striking down Clean Power Plan.

Senate passes the Inflation Reduction Act (51–50), containing $369 billion in funding for climate and clean energy measures.

CLIMATE CHANGE LAW

SECOND EDITION

Chapter 1

INTRODUCTION

Every day, it seems, climate change is in the news. We are constantly reading about unprecedented droughts, floods, or heatwaves linked to climate change. We read about new climate initiatives and laments about the narrow window for escaping disaster. Some days we read that another country is expanding its use of coal or that it is investing heavily in renewable energy. And this is not to mention the ongoing political debates, and the court rulings that make headlines now and then. It can be exhausting to try to keep up—as we know well, since it's our job to do so.

This book is about one crucial element of the climate crisis, the response of the legal system to the ongoing fundamental transformation of our planet. When people think about fighting climate change, they think of scientists and engineers, not lawyers. But law has been central to the effort to deal with climate change. The legal response has been multi-leveled, from the state and local level up to the level of international negotiations. It has also been multi-dimensional, ranging from the use of carbon taxes to plans to adjust to impending changes like sea level rise. Institutionally, international organizations, state and local governments, the Supreme Court, the EPA, and even banks, corporations, and other private actors have all played major roles.

Thirty or forty years ago, only a few scientists or environmentalists might have seen the need for a book on climate change. Even 20 years ago, there would have seemed little need for a book on law and climate change—apart from a couple of international agreements, there just wasn't that much law to talk about. But today, the situation is quite different.

At all levels, from cities to the global negotiations involving nearly two hundred countries, laws, regulations, and court decisions relating to climate change have burgeoned. They cover topics ranging from international finance mechanisms to alleviating the physical impacts of climate change to regulations of the electrical grid to carbon trading systems to aviation emissions.

To get a sense of how varied these legal measures are, consider some of the following examples:

- The European Union and the state of California both created ambitious cap and trade programs to limit carbon emissions, while British Columbia has imposed a carbon tax.

- The U.S. EPA has imposed regulations on emissions of methane (a potent greenhouse gas) from the oil and gas industry, along with limits on carbon dioxide emissions from cars.
- The Supreme Court ruled that a state had standing to challenge EPA's refusal to regulate greenhouse gases. Fifteen years later, it ruled that EPA had gone too far in its efforts in regulating emissions from electric power plants.
- An international agreement signed by nearly every country in the world required each country to set its own target for regulating carbon.
- A dozen states have set targets for achieving net-zero emissions. The Ninth Circuit rejected a federalism challenge to a state law regulating the carbon emissions for particular fuels.
- Many U.S. states and cities, dozens of the world's poorest countries, and many more developing and industrialized countries have developed national adaptation plans or programs.

These are merely samples of hundreds of legal developments in the past fifteen years.

The legal effort to climate change has not gone unquestioned. Far from it. The gains that have been made have been hard fought, and the battles continue. The Trump Administration attempted to eliminate every major federal climate regulation and many that were not so major. The Biden Administration is now painfully attempting to undo those actions. Trump himself once described climate change as a Chinese hoax designed to ruin the U.S. economy. As one legacy of the Trump era, climate actions must confront a sharply more conservative judiciary, with many judges and Justices who tend to be skeptical of regulation in nearly all forms.

This book is intended to introduce readers to the evolving legal framework relating to climate change. This chapter sets the scene. We will begin with a discussion of the science of climate change (a subject we will return to in the following chapter). Scientific confidence has grown over the past few decades about the reality of climate change, the role of greenhouse gases, and the present and future harmful impacts. That scientific knowledge provides the foundation for legal and policy efforts at the domestic and international level. Although there are areas of uncertainty, the basic facts about anthropogenic climate change are firmly established.

Today, very few climate scientists doubt that the climate is changing and that human activities are the primary cause.

After laying this scientific foundation, we will consider some of the barriers to creating an effective response to potential climate risks. Because greenhouse gases are effectively mixed in the global atmosphere, no single country can either limit the process of climate change or protect itself against the impacts of climate change. Instead, a planetary effort is required. Obtaining global cooperation is never easy, and in this case, it is further hindered by uncertainties about the timing and extent of harm, our general lack of experience with problems having multi-century footprints, and uncertainty about how to decarbonize our energy systems while continuing to allow economic development. Some key countries have powerful fossil fuel sectors, not least the United States. Given the obstacles, it seems somewhat surprising that it has been possible to mount any legal response to climate change, even though the effort so far falls short of what is needed to achieve the globally agreed goal of holding the increase in the global average temperature to well below 2°C over pre-industrial levels.

The chapter will close with a brief overview of the remainder of the book. The two major prongs of climate law involve reducing emissions to limit future climate change ("mitigation") and taking steps to limit harm from whatever climate change does occur ("adaptation"). We discuss both the U.S. and international dimensions of mitigation and adaptation. Finally, we will discuss geo-engineering as an alternative, or complementary strategy to mitigation—basically, a variety of techniques ranging from mundane to audacious (and arguably dangerous) schemes for cooling the planet after emissions are already in the atmosphere.

We live in tumultuous political times and forecasts about future policy are perilous. But to date, climate change law has shown a remarkable degree of resilience in adapting to barriers. This has been largely due to the range of institutions that have the potential to contribute to climate policy. The complexity of the institutional framework means that progress can continue when one policy channel is blocked. But the same complexity can make policy change complicated and difficult.

I. Climate Change and Its Impacts

The root source of climate change is the emission of greenhouse gases, primarily carbon dioxide (CO_2). These gases have a single common feature: they allow higher-frequency radiation such as visible light to reach the surface of the earth but are opaque to the lower-frequency radiation that is emitted back from the surface of the earth. The upshot is that the same amount of energy goes in, but less of the energy is able to escape. This energy imbalance causes warming.

The physics of this "greenhouse effect" is well-understood and undisputed. Where things get more complicated is the feedback resulting from this added energy. For instance, warming melts ice, leaving ground or ocean exposed. These areas are darker than ice and heat up more in the sun, causing additional ice to heat and beginning the cycle all over. Equally, the greater heat evaporates more water. This is likely to trap even more heat; but, if it results in greater cloud cover during the day, sunlight is reflected back into space, causing a cooling effect. Scientists use enormously complicated computer models to sort through these feedbacks. But through a combination of modeling and increasingly detailed data about the composition of the earth's atmosphere across time and the planet's temperature and precipitation record, scientists are now virtually certain that the climate is changing, that the change is caused by emissions of greenhouse gases, and that continued emissions will result in additional warming. We discuss some of the scientific issues, and the remaining gaps.

Scientists are now able to say with confidence that increasing concentrations of greenhouse gases in the atmosphere will cause significant warming and other dangerous climatic changes. These conclusions are not the results of studies by a few over-confident scientists. Instead, they reflect hundreds if not thousands of studies by scientists all over the world. Some critics have accused scientists of engaging in groupthink, but to anyone reading the studies, it is impressive how much effort goes into validating results and testing alternative explanations.

Not only are scientists confident that climate change will occur in the future, but they are also confident that it has already begun. We will dive more deeply into the findings of climate scientists in the next chapter. Here, we need only to summarize a few key points. Once climate change could be seen as a far distant threat decades in the future. That has long since stopped being true. Each recent decade has had higher average temperatures than the last. This change has been accompanied by an escalation of extreme events. We also know that further emissions of the climate change will make things much worse as we cross temperature thresholds at 1.5°C and then 2.0°C. There are still areas of uncertainty, such as how quickly species loss will mount with higher temperatures, or just how sensitive the climate is to emissions. But we do know the basic facts with more than 95% certainty—the level of confidence that is normally translated by lawyers into "beyond a reasonable doubt."

We also know that society will face great challenges in adapting to whatever future changes we are unable to avid. So will ecosystems. Certain ecosystems are particularly sensitive to climate change. Tropical species, for example, tend to have narrow ranges of temperature tolerance, which is simply an indication of how intense competitive pressures have made many species very specialized within particular niches. The narrower temperature ranges for tropical organisms may make them more vulnerable to climate change when they are unable to move far enough or quickly enough to maintain their preferred temperature ranges. Tropical reefs are biodiversity hotspots, but the reefs themselves are already threatened as waters become warmer and more acidic.

The severity of these changes depends critically on the future trajectory of greenhouse gas emissions. Most of the book will focus on efforts to control future emissions in order to prevent or at least moderate further climate change. We will end, however, by discussing two other issues: what kinds of legal responses we can draw upon to cope with a changing climate and what legal response might be needed to encourage and regulate efforts to limit the effects of climate by, for example, sucking CO_2 out of the atmosphere or even cooling the planet by reflecting more sunlight back into space.

II. The Policy Challenges

Climate change is a big problem and that alone would make it a difficult one. It has some special features, however, that make climate action all the more daunting. Addressing climate change requires efforts on a global scale. Humans don't have a strong track record for international cooperation. There is also much that remains uncertain about climate change, complicating the job of mapping out the best policies. And finally, the long-term consequences of climate change require us to think about future generations to a degree humans do not find natural, and law struggles to reflect.

A. The Collective Action Problem

Cooperation encounters inherent difficulties. A classic example is a problem that faced many medieval and early modern villages. Those villages had a "commons" where all of the peasants were entitled to graze their animals. (To this day, Boston has a "commons" that now functions as a park.) This history became the basis for an important approach to conceptualizing environmental problems. In a classic article,[1] Garrett Hardin explained what he called the tragedy of the commons and analyzed the economic logic of the commons. "Rationally," according to Hardin, "each herdsman seeks to graze as

---

[1] Garrett Hardin, *The Tragedy of the Commons*, 162 SCIENCE 1243, 1244–1245 (1968).

many cattle as possible on the commons, because he gets the full benefit of selling each additional animal but suffers only a fraction of the harm to the pasture caused by the additional grazing. The result is individually rational but collectively tragic." As Hardin said:

> Each man is locked into a system that compels him to increase his herd without limit in a world that is limited. Ruin is the destination toward which all men rush, each pursuing his own best interest in a society that believes in the freedom of the commons. Freedom in a commons brings ruin to all.[2]

Hardin observed that pollution is a kind of "reverse" commons issue, in that it is individually rational for each individual to avoid the personal cost of pollution control, which leads to the collectively undesirable outcome of "fouling our own nest," so long as we behave only as "independent, rational, free-enterprisers."[3] Hardin's depiction of the tragedy of the commons has been critiqued as offering a simplified version of the challenges involved in regulating the commons. For example, in her Nobel prize winning work, economist Elinor Ostrom showed how management of the commons is possible and occurs in many places under well-defined circumstances. Ostrom's work challenges Hardin's view of the tragedy of the commons, but it also demonstrates how complex it is to regulate the commons and, in particular, to respond to climate change.[4]

In its many different forms, the tragedy of the commons can be considered a special case of what game theorists call a prisoner's dilemma. This scenario gets its name from the following illustrative story. Consider two prisoners, charged with being involved in the same crime but held in different rooms. If neither one confesses, the prosecutor will have to let them both off with a minor charge. If both confess, they will get a heavier sentence. But if one confesses and the other does not, the one who confesses will be acquitted, whereas the other one will have an extra heavy penalty. Whatever the other party does, it is always best to confess. If the other party confesses, then it pays to confess to avoid getting the extra-heavy sentence for remaining silent. If the other party doesn't confess, then it pays to confess so as to get the acquittal. Using the same logic, both parties

---

[2] *Id.* at 1244.

[3] *Id.* at 1245.

[4] Elinor Ostrom, GOVERNING THE COMMONS (1990); Elinor Ostrom, *A Polycentric Approach for Coping with Climate Change* 39 (World Bank Policy Research Working Paper No. 5095, 2009), http://documents.worldbank.org/curated/en/480171468315567893/pdf/WPS5095.pdf [https://perma.cc/72FK-4BJV] (Ostrom suggests that there are no "optimal" solutions and "encourages experimental efforts at multiple levels."

confess, even though this leaves them worse off than in the situation where they both stay silent. So, just as with the tragedy of the commons the individually rational set of actions leads to an inferior outcome for both of them.

Climate change is an apt illustration of this dilemma. If every country cuts emissions, they can collectively limit climate change, benefitting everyone. But it is always in the interest of any one country to continue emitting, allowing it to free-ride on the efforts of others. If the country cuts emissions, after all, it will only receive a small amount of the benefit while incurring the full costs of those cuts. So, the rational strategy for each country is to do nothing to cut emissions, meaning the climate change goes unchecked and all are worse off.

Yet, all countries would be better off if they could somehow make an enforceable deal to cooperate. But such deals are difficult to negotiate and enforce on the international level because the prisoner's dilemma takes hold again: each individual country has an incentive to drag its feet in the negotiations rather than assume burdensome responsibilities; once a deal is made, each country has an incentive to breach the agreement and allow others to bear the costs of carrying it out.

The upshot is that, in the absence of some mechanism for making and enforcing cooperative agreements, each nation may find inaction to be the most sensible individual choice, even though everyone also knows that collective inaction will only lead to disaster. Within individual nations, the central government can lead the way out of prisoner's dilemmas by forcing a cooperative solution. In the international sphere, the problem is more difficult, but as we will see below, some progress has been made.

The prisoner's dilemma poses a considerable barrier to international environmental regulation. Indeed, given the formidable barriers, we do not seem to have a good explanation for why international environmental regulation has been implemented *at all*. Smaller-scale prisoner's dilemmas also occur at the national level. Fortunately, at their best, humans seem to be able to cooperate for the common good, even when some economic theory predicts failure. But there is nothing easy or inevitable about the process.

The collective action problems are truly daunting: climate change requires a coordinated response involving every country (or at least every high emitting country) on earth, a task made much harder because of the huge differences in development across the globe and the large economic interests involved in fossil fuels.

Despite these difficulties, as we will see in later chapters, an increasing number of individual countries and even state governments have taken steps to cut emissions, and there has also been progress toward enhanced international cooperation. This degree of progress is all the more remarkable because, as discussed below, the collective action problem is only one of the barriers to effective action.

In this sense, humanity has done better than one might have expected to meet the climate crisis. Even so, it remains to be seen whether we will act forcefully and quickly enough to avoid disastrous consequences.

B. Uncertainty

As two leading climate scientists have said, "[t]he further we push our Earth outside of its mode of operation of the past millennia, the further we steer it into uncharted waters."[5] Climate policy must contend with this uncertainty. We will discuss the nature of the uncertainties in more detail in the next chapter, but a quick preview is helpful at this point.

The problem is that despite very sophisticated and extensive efforts by scientists, the residual areas of uncertainty remain substantial. Of course, it is completely clear that the earth has been warming: scientists say the evidence for warming is unequivocal, marked by many observed changes in physical and biological systems. But the details are subject to various shades of uncertainty.

The effects of climate change depend on how much carbon we emit and on the sensitivity of the climate to greenhouse gases. This sensitivity is measured by determining how the climate would respond to a permanent doubling of atmospheric CO_2. The problem is that scientists haven't been able to determine this sensitivity precisely. The range of uncertainty has narrowed, but there is still some uncertainty in translating emissions levels into temperature levels. There is greater uncertainty about how different global temperatures translate into local temperature and precipitation, and even more uncertainty about ecological and social responses.

The future trajectory of emissions is also an unknown: it depends on future human actions. Scientists deal with this problem by using a variety of scenarios. The difference between these scenarios is substantial, but there are also uncertainties within each

---

[5] DAVID ARCHER AND STEFAN RAHMSTORF, THE CLIMATE CRISIS: AN INTRODUCTORY GUIDE TO CLIMATE CHANGE 152 (2010). In particular, the past seven thousand years during which civilization has arisen were unusually stable, so climate change is well outside the circumstances faced by groups within the historical record. *See* WILLIAM NORDHAUS, THE CLIMATE CASINO: RISK, UNCERTAINTY, AND ECONOMICS FOR A WARMING WORLD 51 (2013).

one.[6] For instance, with an extremely rigorous climate policy, we could keep the temperature increase by the end of the century to 1.0–1.8°C, attaining the target of the Paris Agreement. Completely uncontrolled deforestation and use of fossil fuels could result in a catastrophic 3.7–5.7°C of warming by 2100, while an intermediate scenario would produce very serious impacts at 2.1–3.5 degrees. The differences between scenarios are the most striking, and they provide a strong incentive for climate action. But the range of uncertainty within any given scenario is too big to ignore, especially in planning adaptation efforts.

This uncertainty can make constructive response to climate change more difficult. It is always possible for the fossil fuel industry and politicians to take refuge in the lack of certainty by arguing for further delay. And even among those who are committed to taking action against climate change, lack of certainty makes it harder to agree on the appropriate pace and level of response.

C. Intergenerational Impacts

One distinguishing feature of climate change is its timespan. Carbon emissions remain in the atmosphere contributing to warming for periods up to several centuries. Moreover, once it is warmed, the ocean continues to store heat for a long period of time because of its size. Some impacts on climate patterns could persist even if once warming ends. The result is that the full effects of climate change will play out over centuries, not decades. Many of the people who will be most harmed by climate changes have not been born yet—in fact, their grandparents may not have been born yet.

The intergenerational effects raise knotty philosophical issues. Do living people have a duty to future generations? If so, what is the duty: A duty to ensure that they have at least a minimum level of social welfare? A duty to protect them against physical risks? A duty to preserve the natural world for their benefit? Or a duty to maximize their welfare? All of these formulations are possible interpretations of the concept of duties to future generations, referred to in international law as the principle of intergenerational equity.

Apart from these philosophical issues, intergenerational effects mean that many victims of climate change cannot vote or take part in current climate debates. They are truly unrepresented in the political process and dependent on those now living to take their

---

[6] VALÉRIE MASSON-DELMOTTE ET AL., IPCC, 2021: SUMMARY FOR POLICYMAKERS. IN IPCC, CLIMATE CHANGE 2021: THE PHYSICAL SCIENCE BASIS. CONTRIBUTION OF WORKING GROUP I TO THE SIXTH ASSESSMENT REPORT OF THE INTERGOVERNMENTAL PANEL ON CLIMATE CHANGE 14 (2021), doi:10.1017/9781009157896.001.

interests into account. Political systems do not necessarily encourage politicians to look even a decade ahead, let alone a century.

As we have seen, the barriers to climate action are formidable: the need for global collective action despite the incentive to free-ride on the efforts of others, the uncertainties that provide excuse for delay or indifference, and future generation's political powerlessness though they are the most impacted. And on top of those are self-interests due to vested interests in fossil fuels, and the vicissitudes of politics. Given these barriers, it is not surprising that progress has been slow and halting; what is perhaps surprising is that there has been any progress at all.

Some of the progress can be explained by the incidental benefits of climate action. Actions to reduce emissions can have other benefits besides reducing warming, such as developing renewable industries, reducing dependence on foreign suppliers of fossil fuels, and reducing conventional air pollution. These side-benefits help ease the way. But at the end of the day, the simple explanation is that many people have been motivated to fight climate change because of the simple belief that it's the right thing to do.

III. Policy Responses

Despite the barriers, the world has seen a surge in efforts to address climate change in the past thirty years. Choosing a starting point is a bit arbitrary, but the best time to start the story is probably 1992. That year, the international community adopted the United Nations Framework Convention on Climate Change (UNFCCC). The United States, under the leadership of President George H.W. Bush, signed and ratified this agreement. The UNFCCC acknowledges the seriousness of climate change and calls for a joint multilateral effort to address it. Developed countries such as the U.S. agreed in principle to limit their emissions in order to prevent dangerous levels of carbon in the atmosphere, but the UNFCCC did not contain specific obligations or measures to implement this commitment. The treaty was intended as an expression of general commitment and a framework for further negotiations. Those negotiations have had many frustrations but also some genuine achievements. We will mention only some highlights here, leaving details to a later chapter.

Following the adoption of the UNFCCC, the next step was the negotiation of the Kyoto Protocol in 1997. The Kyoto Protocol obligated developed countries to collectively reduce their greenhouse gas emissions to 5 percent below 1990 levels by the time of the first compliance period from 2008 to 2012. The Protocol embraced market mechanisms, including emissions trading, as a way to reach this goal. The U.S. had insisted on the inclusion of these mechanisms but then failed to ratify the agreement because of concerns about the cost of limiting emissions and about the fact that rapidly developing countries like China were not obligated to limit their emission levels. Many of the parties to the Protocol agreed to a second commitment period from 2013 to 2020. But the Protocol covered too few of the world's major emitters to offer an effective, long-term solution to the problem of climate change.

The world changed dramatically between the adoption of the Kyoto Protocol and the 2015 adoption of its successor, the Paris Agreement. The rise of China as an economic superpower and as the largest net greenhouse gas emitter is probably the most notable change. By 2015, the largest global greenhouse gas emitters included not only the major developed countries, but also a handful of the world's rapidly developing economies. As a result, it was clear that the participation of the United States and China, as the two largest global greenhouse gas emitters was critical to any effort to limit climate change. Equally, it was clear that any agreement must include all of the world's largest emitters as well as having the support of the countries most vulnerable to climate change. Those countries most susceptible to climate change are often the poorest countries as well as islands that faced the threat of disappearing under the sea. Negotiations teetered along precariously until 2015. The biggest issues that divided countries related to equity. Developing countries insisting that developed countries had done the most to create the pool of carbon now in the atmosphere and should bear primary responsibility for fixing the problem.

In December of 2015,[7] the Parties to the UNFCCC successfully negotiated a new successor agreement to the Kyoto Protocol. This new agreement, the Paris Agreement, represent a new, bottom-up, cooperative model for international cooperation. The Paris Agreement commits the Parties to "holding the increase in the global average temperature to well below 2°C above pre-industrial levels and pursuing efforts to limit the temperature increase to 1.5°C above pre-industrial levels," as well as to pursuing complementary efforts with respect to issues such as adaptation and climate financing. The Paris Agreement, however, does not set an overall emission reduction target that is shared among the Parties to the Agreement. Instead, each country must submit a Nationally Determined Contributions (NDC) that reflects the Parties' highest possible ambition within the common but differentiated responsibilities and respective capacities (CBDR) framework. That is, the country must say what it is willing to do to address climate change and explain why its commitment is fair and ambitious. With each subsequent round of NDC

---

[7] The Paris Agreement entered into force on November 4, 2016. As of June 2017, 148 Parties, of the 197 Parties to the UNFCCC, have ratified the Paris Agreement.

submissions, countries must then adopt progressively ambitious goals.

The election of President Trump delivered a blow to the Paris Agreement. Trump officially withdrew from the agreement. Fortunately, no other country followed suit, and the U.S. itself rejoined as soon as President Biden took office.

Given the dynamics of global negotiations and the ongoing uncertainty it breeds, you might expect that individual jurisdictions would continue unabated emissions pending some binding international agreement. That has not proved to be true. In fact, even before the Paris Agreement, a diverse and growing number of governments were stepping up to take leadership roles in addressing different aspects of climate change.

Climate efforts can take place at various levels of government. In the United States, the federal government has had an up and down journey, with forward movement under Democratic presidents and retreat under Republican presidents. At least so far, the retreats have not fully erased the advances. Even Trump left some climate policies untouched, including EPA's authority to regulate greenhouse gases under the Clean Air Act. Many regulatory rollbacks were not finalized, and others are being reversed by the Biden Administration. Some of the federal government's most significant actions have taken the form of major financial support for the development and deployment of renewable energy, first under Obama's stimulus bill and later under Biden's infrastructure bill.

Generally, we expect state and local governments to focus on issues closer to home, leaving international problems to the federal government. But states, provinces, and cities in many countries have risen to the challenge recognizing the local and regional challenges climate change poses for their constituencies. California has been a global leader, with regulations on emissions from new cars, strict caps on utility use of fossil fuels, and its own cap and trade system for carbon emissions. But other states have caught up rapidly, with a growing number of states adopting ambitious targets to reach carbon neutrality and increasing vigorous climate-focused regulatory actions. Even in states where "climate change" is still a politically dangerous thing to discuss in politics, there is often widespread support for renewable energy. Texas has been long controlled by conservative Republicans, but it is also the country's number one source of wind power.

IV. The Framework Governing Climate Action: A Primer

Since not all readers may be familiar with constitutional and administrative law, it may be helpful to have a quick introductory overview. We will go deeper as the occasion arises in later chapters. Readers who are familiar with these areas of the law should skip ahead.

It is best to begin at the constitutional level. Article I of the Constitution gives Congress the "legislative power" and provides a lengthy list of specific topics on which Congress can legislate. The most significant of these is the power to regulate interstate commerce. Although the exact scope of this power has been controversial through most of our history, today it is interpreted to include any commercial activity, whether or not the activity crosses state lines. Congress has yet to pass legislation explicitly regulating greenhouse gases, but it did pass a bevy of other environmental statutes in the 1970s based on the commerce power. These statutes have been deployed to address issues dealing with climate change, particularly during the Obama Administration.

Article II of the Constitution addresses the powers of the President. In the international sphere, the President has powers that are independent of Congress. For instance, this includes the sole authority to decide whether or not to recognize foreign nations. The President also has the power to make treaties, but treaties must be ratified by the Senate to be effective as a matter of U.S. domestic law. Presidents have also made hundreds of "executive agreements" with foreign nations, which have not been submitted to the Senate for ratification (including the Paris Agreement). This is another controversial area of U.S. constitutional law because there is no consensus about what agreements can be made by the President independently and what agreements must be submitted to the Senate for approval.

The President's powers with regard to domestic policy are more limited, at least as a formal constitutional matter. Because Article II grants the President the "executive power" and responsibility for administering the laws, the President has some power to control the internal operations of the executive branch. There is also an ill-defined zone in which the President can take legally binding actions where Congress has been silent. But in the domestic sphere, such actions are unusual. The President's primary legal role domestically is to oversee the implementation of congressional enactments. Indeed, the Constitution imposes a duty on the President to "take care that the laws be faithfully executed."

One of the key instruments of presidential control over the execution of the law is the appointment of high-level officers within the executive branch. The President has the power to appoint several hundred officials, with Senate confirmation required for the higher-level appointees. The general rule is that the President has the power to terminate government officers at will. There are exceptions for so-called "independent agencies," but the legitimacy and scope of these exceptions has remained controversial since the early Twentieth Century. The Federal Energy Regulatory Commission (FERC) is one of those independent agencies, but the Environmental Protection Agency (EPA) is not. The Administrator of EPA is removable at will by the President, although this has not stopped some Administrators from pushing hard to make sure the agency's voice is heard.

Federal judges are appointed by the President with the "advice and consent" of the Senate, but once appointed, they serve for life. This gives them an unusual degree of independence. The federal courts have jurisdiction over cases arising under federal law. Historically, one of their most important roles has been judicial review of the constitutionality of federal and state laws. But in the climate arena, their most significant role has been reviewing the legality of actions by EPA and other agencies implementing federal statutes. That role takes place within the framework of administrative law, which we will discuss below.

Because the U.S. has a federalist constitutional system, states retain an important role in the legal system—and, as we will see, in creating climate change policy. States are considered to have inherent lawmaking power, so they do not point to any source of authority within the federal constitution as a basis for enacting laws. But their powers are subject to limitations under the federal constitution. The Fourteenth Amendment requires them to respect individual rights; the scope of those rights has been defined in a series of Supreme Court decisions over many decades. In terms of climate law, however, the main issue is usually whether state activities go beyond their role in the federalist system.

Challenges to state laws can be based on two legal doctrines known as the dormant commerce clause and federal preemption. The mysterious sounding "dormant commerce clause" puts limits on state interference with interstate commerce. Dormant commerce clause cases involve interference with interstate commerce. To simplify greatly, a law will be invalidated if it appears to discriminate against out-of-state firms or if the state has only a slender justification for a serious burden on commerce. Federal preemption maintains the supremacy of federal law. Federal laws trump state laws if the state laws conflict with federal laws or interfere with their operation. There are a lot of federal laws and consequently a lot of different sources for possible preemption claims. For our purposes, the most significant preemption problems tend to involve federal regulation of some aspects of the electric power system and federal regulation of emissions from new vehicles.

As mentioned above, the most important role of the courts in climate change law is reviewing the actions of federal agencies like the EPA. The body of rules surrounding this review is known as administrative law. If you're a law student who is at all interested in environmental law or climate change, you need to take that course! If you haven't done so yet, here is enough of an outline to get you through the rest of this book: Under the Administrative Procedure Act (APA),[8] there is a set of procedures that has to be followed for almost all regulations. The initial step is the publication of a notice in the Federal Register describing a proposed rule along with a detailed explanation of why the agency believes the rule is required. The agency must then allow interested parties an "opportunity to comment" for at least thirty days. Then, the agency must issue, in conjunction with its promulgation of the rule, a "concise general statement" of the rule's basis and purpose. (There are a limited number of exceptions to these requirements.) The agency needs detailed evidence to support its decision and must respond clearly to any significant arguments made by opponents of its proposal. It is not required by Congress, but for the past forty years presidents have also required agencies to prepare a cost-benefit analysis for every major rule. All of this adds up to a slow, cumbersome rule-making process. Hopefully the result of the effort is better rules and increased transparency. Still, no one can claim that the regulatory process is nimble. It can be difficult for a presidential administration to complete a major rule and litigate its validity before that President leaves office, creating the risk that the new President may choose to simply restart the whole process.

The extent of judicial review generally depends on the kind of issues involved in the case. In particular, it is useful to distinguish between issues relating to pure questions of law, issues involving the factual basis for the agency's action, and issues relating to the agency's procedures.

The first kind of issue involves only a question of law, such as what the word "system" means in a particular section of the Clean Air Act. That sounds like a minor technicality, but in fact it determined the fate of the Obama Administration's signature climate change regulation, the Clean Power Plan. The Supreme Court's approach to reviewing an agency's interpretations of statutes are currently in flux. The governing legal doctrines are too complex to

---

[8] 5 U.S.C. §§ 551–559.

cover in depth, but here is the crux of current doctrine. First, if a case presents an issue of sufficient importance (what the Supreme Court now calls a major question), the agency must show that it has clear authority from Congress to back its decision. Second, if there's no major question, the court will decide the statute's meaning for itself if it considers the language to be clear. Finally, if the agency gets past that hurdle, there's a reward: the Court will accept any reasonable interpretation the agency chooses. This last step is called the *Chevron* doctrine after the case where it originated.[9] Conservative judges don't like it; and no one know how long it will stay on the books. All of this is, of course, a simplification, which is why administrative law needs a separate course. But it should be enough to get you through this book.

The second kind of issue involves the agency's policy choices. Judges are not supposed to be in charge of deciding energy and environmental policy. The relevant agency (e.g., the EPA and FERC) has more expertise, and the heads of energy and environmental agencies are appointed by (and mostly removable) the president, which gives them a democratic pedigree that courts lack. In reviewing an agency's policy decisions and scientific determinations, courts apply what's called the "arbitrary and capricious" test. The name makes it sound like any halfway plausible support for the agency's action will do, but in practice the test can be much more rigorous. The basic idea is that, at the time it acted, the agency had to give a reasoned explanation for what it is doing based on the evidence before it at the time. The court's role is to decide whether the agency considered all of the relevant factors and identified evidence in support of the decision.

V. Roadmap

Our central theme is the difficulty of dealing with a massive global problem in the context of a fragmented governance system. But the counterpoint is that dedication, entrepreneurship, and ingenuity have done much to overcome the difficulties.

Before we launch into the details, it is helpful to have an overview of the remainder of the book. Following this introductory chapter, Chapter 2 will focus on climate science and policy. Climate law has been driven by advances in scientific understanding. Yet, although we know a great deal about the likely trajectory of climate change, there are still significant uncertainties about its severity and about how global changes will translate in terms of local impacts. Chapter 2 discusses the scientific basis for our knowledge of climate change and explains the extent of remaining uncertainties.

---

[9] Chevron U.S.A., Inc. v. NRDC, 467 U.S. 837 (1984).

Chapter 2 then turns to a discussion of the policy issues. There are really three major areas of debate. First, how do we account for harms to ecosystems and biodiversity? Second, how should impacts on future generations be taken into account? Third, what about the distributional impacts of climate change? Economists have proposed techniques for dealing with those issues, and we will devote particular attention to the issue of future generations, which despite narratives to the contrary, are not merely theoretical. The federal government, for example, has grappled with these issues in its controversial effort to determine the social cost of carbon, an important metric in setting emissions policy.

With Chapter 3, we turn to the legal framework for climate policy. At the international level, the United Nations Framework Convention on Climate Change (UNFCCC), the Kyoto Protocol, and the Paris Agreement establish the legal framework for international efforts to address the causes and consequences of global climate change. Based on the goal of "stabilizing atmospheric concentrations of greenhouse gas at a level that would prevent dangerous anthropogenic interference in the climate system", the UNFCCC is the primary forum for ongoing multilateral negotiations on climate change. Participation in the UNFCCC is nearly universal, with 197 parties—representing all UN Member States—having ratified the treaty. However, despite widespread participation and nearly 30 years of negotiations, the international regime is plagued by tensions and by an ongoing inability to structure an effective legal foundation for collective efforts to address climate change.

Within this context, Chapter 3 begins by providing a brief overview of the history of the international climate change regime. The chapter explores the history of the regime with a view towards explaining underlying legal principles, such as common but differentiated responsibilities, and exploring how evolving understanding of the problem of climate change creates new challenges to interpreting and applying these principles.

Chapter 3 then introduces the innovative set of tools that the Kyoto Protocol recognizes as central to achieving economically efficient emissions reductions, including emissions trading, the clean development mechanism, and joint implementation. Here, we will lay the foundation for more detailed discussions of sub-global mitigation efforts in the chapters that follow.

Chapter 3 also explores the role of the UNFCCC moving forward. This Chapter will introduce the primary points of tensions underlying ongoing discussions and examine the different roles that traditional multilateral legal institutions can and should play in this climate context. The Paris Agreement provides a platform for further discussions, but the strength of this trajectory remains to be seen.

In Chapter 4, we probe more deeply into the economic instruments that have been at the core of climate policy since the negotiation of the Kyoto Agreement. One approach to reducing carbon emissions (and thereby limiting the scope of future climate change) is to price carbon emissions, either through a carbon tax or through a carbon trading system. Chapter 4 begins with a brief discussion of the policy issues relating to these tools. These include arguments for preferring a tax to a trading system or vice versa, as well as arguments for using non-market instruments.

Chapter 4 then discusses experience with implementing carbon-trading systems. The EU trading systems has the most extensive history. But state emissions trading system, especially the RGGI system in the East and AB 32 in California, are also important trading efforts. Experience has shown that the workability of any emissions trading system depends heavily on the details of its structure. Chapter 4 also discusses one of the most controversial questions relating to trading systems: the role of credits and offsets. We will introduce key issues such as enforceability and additionality. The experience under the Kyoto CDM provides some warnings about the pitfalls. We will also discuss the REDD+ approach to forest preservation as a possible source of credits.

Chapter 5 takes a closer look at methods for limiting emissions from the energy sector, focusing on the two largest sources of domestic greenhouse gas emissions, the electric power and transportation sectors. Electricity regulation, in particular, is a critical component to mitigation strategies since electric power generation is the second largest source of greenhouse gas emissions in the United States (following transportation). Electricity regulation is a complex subject; here we will only cover some of the features most relevant to carbon reduction. We will discuss how regulation of the electric power system is in flux, with recent efforts focusing on increasing competition and energy diversification in the electric power market. We will also discuss how efforts at national level to create a smarter, cleaner, and more efficient electric power grid are complemented by efforts at the state level to encourage the growth of renewable energy. To this end, we will explore steps being taken by the states to incentive renewable energy, focusing on the widespread use of renewable portfolio standards in U.S. states. We will also probe the legal problems involved in siting new transmission lines, windmills, and solar installations. The relationship between federal and state regulation of electricity markets is particularly intricate and will not be treated in detail, but some of FERC's key initiatives will also be discussed.

Transportation is currently the largest source of U.S. greenhouse gas emissions. Biofuel policies, both at the state and federal level, have attempted to address transportation emissions. We will discuss the federal legislation and the leading state effort, California's Low Carbon Fuel Standard (LCSF). We will also discuss federal fuel efficiency standards, which have given rise to litigation in the context of climate change, as well as various other efforts at the state level to reduce the footprint of the transportation sector, including through efforts to promote zero emission vehicles and land-use change and city planning decisions designed to minimize the total number of miles traveled each year. Finally, we will briefly discuss efforts to curb emissions from the international aviation and shipping sectors.

With Chapter 6, we turn to the highly contentious issue of federal climate policy. There was a strong effort during the Obama Administration to pass federal climate legislation. While the pathbreaking 2010 Waxman-Markey bill cleared the House, it failed to clear the Senate and was then doomed by a shift in party control in the next election. In the dozen years since then, Congress has been deadlocked. It has passed some measures to promote renewable energy financially, and it passed one last bill to deal with a particularly nasty group of chemicals that cause warming. Still, in terms of regulating the most prevalent greenhouse gases, carbon dioxide and methane, Congress has left a policy vacuum. For this reason, federal agencies such as EPA have had to rely on existing statutes dealing with more general problems such as air pollution, rather than on more specific climate legislation.

At the federal level, the Supreme Court has assigned the task of reducing carbon emissions to EPA. In *Massachusetts v. EPA*, the Court upheld a challenge by state governments and others to EPA's refusal to regulate greenhouse gases under the Clean Air Act. Then, in the *AEP* case, it held that EPA's regulatory authority "displaced" the federal common law of nuisance. The upshot was that EPA has become the primary forum for climate policy, a role it resisted under Bush, embraced under Obama, attempted to repudiate under Trump, and is now reclaiming under Biden.

The Clean Air Act is an extremely complex statute, and EPA has turned to some less-used provisions of the statute as tools for regulating carbon emissions. We will explain why EPA has turned away from the core provisions of the statute, such as the use of state implementation plans. One set of regulations by EPA, issued under the "PSD" portion of the statute, was largely upheld by the Supreme Court in the *UARG* case. We will devote careful attention to that opinion.

EPA then turned to section 111 of the statute to regulate existing power plants. In the past, this section had been used almost exclusively to regulate new power plants. After issuing regulations governing new plants, EPA also proposed sweeping regulations of existing plants under the previously obscure subsection 111(d). EPA's initial efforts were rebuffed by the Supreme Court in the case of *West Virginia v. EPA*, and at this writing the agency is considering its options.

Finally, we will discuss the issue of climate standing, which plays an important role in litigation. The Supreme Court upheld a state's claim of standing in *Massachusetts v. EPA*, but there is considerable dispute in the lower courts about the breadth of the holding.

Perhaps surprisingly, efforts to cut carbon are not limited to the federal government. Chapter 7 gives an overview of what states (and localities) have been doing. The other purpose of this chapter is to examine some of the legal issues relating to state regulation. The most important challenges have come under the dormant commerce clause, as exemplified by the Ninth Circuit's decision in the *LCFS* case. But there are also potential issues under foreign affairs preemption and the compact clause. Since statutory preemption claims could arise under diverse statutes, we will only provide a general discussion of the doctrinal framework here. Under the Trump Administration, the ability of state and local governments to take climate action was thrust to the forefront, given the Administration's rejection of federal climate action. But it seems likely that, regardless of who is in power in Washington, the federal government will not catch up with the leading states for years to come.

Because of the amount of carbon already in the atmosphere, some amount of warming is already "baked-into" the climate system. At a minimum, humans will have to cope with that inescapable climate change, and failure to control climate emissions could mean that the changes will be much larger. Climate change means not only rising temperatures, but also sea level rise and more serious heat waves, floods, and droughts. Chapter 8 examines the challenge of adapting to climate change.

Chapter 8 begins with a discussion of current predictions about climate impacts. We will then focus more specifically on the process of adaptation planning and challenges inherent in coordinating adaptation efforts at different levels of governance. We will explore the role that the UNFCCC and other international organizations play in facilitating adaptation planning as well as adaptation financing. To this end, we will explore the evolving world of international climate finance. We will then unpack the relationship between climate change and disaster law and consider how the intersection between climate change and disaster highlights ongoing and urgent adaptation and financing needs.

Chapter 8 will then turn to domestic adaptation to explore the largely facilitative role that the national government has played in adaptation, thus far. We will then explore how particular states and cities are approaching the range of challenges that climate change poses for their citizenry.

Finally, Chapter 8 explores the interesting legal problems that arise at the intersection of adaptation and property rights. Some of the most pressing adaptation challenges involve the impact of sea level rise in coastal areas. In these coastal zones, while it would be desirable to prevent further development in areas that are threatened by rising seas and even to move existing development, perhaps through the use of rolling easements, takings issues threaten such efforts. As a result, adaptation efforts often involve developing creating solutions to challenging legal and physical problems.

Chapter 9 closes the book with a discussion of a very different type of response to climate change: geo-engineering. Geo-engineering techniques are used to reduce the amount of warming resulting from any given level of carbon emissions. The two primary categories of geo-engineering include carbon dioxide removal (CDR) and solar radiation management (SRM). The most frequently cited techniques for CDR include capturing carbon from fossil fuel fired electricity generating units, afforestation, ocean fertilization, and the use of biochar. The most commonly considered SRM techniques involve using mirrors in space or injections of aerosols into the stratosphere to reduce the amount of radiation reaching the surface. Such efforts would likely have significant side effects of various kinds. This is an area where there is great scientific uncertainty and little legal or regulatory guidance, making it ripe for critical inquiry. We will discuss different proposed methods to geoengineer the climate, potential problems under international law, possible liability issues, and ongoing efforts to develop governance and regulatory regimes to control geo-engineering at both the international and domestic levels.

We are keenly aware of how quickly the legal landscape can change in this field. What appeared to be a likely pathway to continual strengthening of U.S. climate policy collapsed unexpectedly with the 2016 U.S. elections, suddenly creating a much more fraught situation. But, however the field develops in the future, human beings are going to have to cope with the physical phenomenon of climate change one way or another. Understanding the current legal framework is the first step in preparing for the future.

Further Readings

BILL GATES, HOW TO AVOID A CLIMATE DISASTER: THE SOLUTIONS WE HAVE AND THE BREAKTHROUGHS WE NEED (2021).

NAOMI KLEIN, THIS CHANGES EVERYTHING: CAPITALISM VS. THE CLIMATE (2015).

MICHAEL E. MANN, THE NEW CLIMATE WAR: THE FIGHT TO TAKE BACK OUR PLANET (2021).

JOSEPH ROMM, CLIMATE CHANGE: WHAT EVERYONE NEEDS TO KNOW (2015).

GERNOT WAGNER AND MARTIN L. WEITZMAN, CLIMATE SHOCK: THE ECONOMIC CONSEQUENCES OF A WARMING PLANET (2016).

Chapter 2

CLIMATE SCIENCE AND ECONOMICS

Absent concerted action, climate change will pose a substantial threat to humanity by the end of this century. As the Intergovernmental Panel on Climate Change (IPCC) warned in 2014:

> Without additional mitigation efforts beyond those in place today, and even with adaptation, warming by the end of the 21st century will lead to high to very high risk of severe, widespread, and irreversible impacts globally. (In most scenarios without additional mitigation efforts . . . , warming is more likely than not to exceed 4°C above pre-industrial levels by 2100. The risks associated with temperatures at or above 4°C include substantial species extinction, global and regional food insecurity, consequential constraints on common human activities, and limited potential for adaptation in some cases.[1]

In this chapter, we will explore the evidence behind these predictions and their policy implications. The first half of the chapter is devoted to climate science. The second half concentrates on climate economists' efforts to determine just how rapidly and stringently emissions should be controlled. We will also consider some of the ethical disputes relating to the economics of climate change.

The material in this chapter serves more than one purpose. It is obviously important to know the reasons behind regulation of climate change before considering the specifics. Beyond that, it is critical to understand the science behind climate change when assessing our progress in confronting the problem and the work that remains to be completed. Understanding the basics of climate science is especially important for issues discussed later in the book regarding adaptation to climate change and geo-engineering. Similarly, the discussion of climate economics will set the stage for our later discussion of alternative mitigation tools such as carbon trading versus taxes. Climate law is still in flux, so it is important not only to understand its current state but also how changes in scientific knowledge and economic analysis may help shape its future.

I. Climate Science

Without the efforts of thousands of scientists around the world, we might not even be aware of how our climate is changing, let alone

---

[1] INTERGOVERNMENTAL PANEL ON CLIMATE CHANGE (IPCC), 2014 SYNTHESIS REPORT: SUMMARY FOR POLICYMAKERS, at 19.

the reasons or the future direction of changes. As everyone knows, weather can vary enormously from day to day or even year to year. It's necessary to filter out these fluctuations to determine what's happening with the climate. Climate differs from weather in that weather refers to atmospheric conditions over short periods of time while climate refers to long-term averages of daily weather. For example, we would generally say that Arizona has a drier climate than Ohio even though there may be days when it rains somewhere in Arizona but not Ohio. To investigate climate change, scientists have had to develop global information about weather, filter out the fluctuations, and then probe deeply into the atmospheric dynamics that shape longer term climate trends.

Every five years, the IPCC publishes a synthesis of the available scientific information. Even that synthesis runs thousands of pages, and it provides the best way of learning more about the technical details. Here, we will forego the details and give the proverbial view from ten thousand feet—an apt phrase in this setting since a good deal of data derives from satellites. This section provides a very brief overview of what scientists have learned. It addresses three key topics: the scientific basis for our knowledge, the sources of emissions, and the areas where our understanding is still incomplete.

A. How Do We Know About Climate Change?

Scientists have many sources of information about the past and present climate. Today, we have reports from weather stations all over the planet and from satellites. Systematic measurements go back about 145 years, but we can go back much further with data from tree rings, sediments in lakes and oceans, and gases trapped in ice cores from glaciers. The evidence shows clearly that the Earth's surface temperature has increased about 1°C over the past century, with much of the increase occurring in the past 35 years. Each of the last four decades has been successively warmer at the Earth's surface than any preceding decade since 1850. According to the IPCC:

> Warming of the climate system is unequivocal, and since the 1950s, many of the observed changes are unprecedented over decades to millennia. The atmosphere and ocean have warmed, the amounts of snow and ice have diminished, and sea level has risen.[2]

There are many signs of this warming. They include melting of glaciers and ice in Greenland and Antarctica, measurable increases in sea level, and changes in the ranges and behavior of animals and plants. Even those who oppose taking any action to deal with climate change generally concede this point today. They argue, however, that

---

[2] *Id.* at 1.

the warming is due to natural causes, not human emissions of greenhouse gases, or at least that greenhouse gas emissions are only a partial cause of climate change. The scientific evidence does not support that view: human activities are overwhelmingly the cause of long-term warming in the past century.

The historic record, going back millions of years, demonstrates a strong relationship between global temperatures and concentrations of greenhouse gases. The basic physics that creates warming has been understood since the late 1800s. We have a much more detailed understanding today, however, because advances in technology have enabled us to capture satellite weather data and to develop sophisticated computer models of the climate system. These models take into account not only the direct warming effect of greenhouse gases but also how that warming works its way through the geophysical system, impacting wind patterns, ocean currents, precipitation, and the cycling of carbon through the biosphere.

The basic idea behind the models is simple. Take a cube of air. Heat, water vapor, and air currents enter and leave through the edges, while solar radiation comes from above during the day and gets reradiated there. Starting at one moment of time, we can predict what will happen to this cube of air in the next moment if we know its current state and those of all the cubes surrounding it. For instance, if this cube is warmer than its neighbors, heat will leak toward the neighbors. There are complex mathematical formulae that can be used to make these calculations. To make projections, all we need to do is to run these calculations for cubes covering the entire earth and its atmosphere, which bring us to the next tick of time (often a half hour), and then repeat the calculations over and over again. Current models also do the same thing with the ocean in order to simulate currents. That's the basic idea, in any event. Actually implementing this idea with thousands of cubes and time spans of decades and centuries is a much more complicated matter. Even with the use of supercomputers, running these models can take extended periods of time.

Climate modeling has developed quickly over the past few decades. The number of scientific publications on climate research has doubled approximately every decade since the middle of the last century. Supercomputer speed has increased by over a million-fold in the past three decades. There has been a shift from using traditional supercomputers (in which all processors use the same memory space) to using a massive number of computers processing in parallel (in which each computer has its own memory space). Computer chip speeds have also grown exponentially, contributing to this shift. These technological advances allow models to be more fine-grained (smaller cells providing more detail on processes) and enable the incorporation of data on ocean currents and other factors too complex to be included in the early models. Using smaller cubes also provides the opportunity to include more details about local topography.

Today, these models are able to include consideration of aerosols (such as sulfur dioxide plumes caused by industrial sources), river and estuary water mixing (affecting ocean salinity), sea ice, and terrestrial processes. Instead of using rough approximations (like average levels of cloudiness in a given locale), models are increasingly able to project cloud formation. Models also increasingly incorporate the relationship between the biosphere and climate, including vegetation changes and soil carbon cycles. Many factors turn out to be relevant: snow-vegetation interactions, evaporation from forest canopies, and soil moisture.

Given that the models are inherently imperfect, what reasons do we have for crediting their results? Perhaps the most fundamental reason is that the cores of the models are based on well-understood laws of physics relating to fluid behavior, thermodynamics, radiation absorption, and other processes. The basic mechanism of the greenhouse effect itself is simple. Greenhouse gases are transparent to visible light, which reaches the surface of the earth and causes warming. The warming results in the emission of infrared radiation, but the greenhouse gases block some of the outgoing infrared radiation, so it's harder for energy to escape into space. Consequently, heat is retained in the atmosphere, ground, and oceans. This physical process is well understood and completely noncontroversial; the complications come from taking into account all the secondary effects of the increased energy retained by the atmosphere and oceans.

Scientists have made strenuous efforts to weed out errors in the data and in their modeling. The models have undergone three important "reality checks." First, some models have been successfully tested for short-term and seasonal weather forecasting, with good results. This provides some grounds for confidence that major weather factors have not been omitted. Second, models have been tested at the component level. Standardized tests are applied to the components through organized activities, such as regularly held workshops on esoteric subjects like "Partial Differential Equations on the Sphere." The physical parameters in the models are tested through case studies, run by programs specializing in cloud systems, atmospheric radiation, and other topics. This is a level of community self-scrutiny that goes beyond the usual mechanism of peer review. Third, models are tested against past and present climate. They have been extensively and successfully used to simulate twentieth century climate changes. That is, if we start a model running from the beginning of the 20th Century, it can provide a good approximation to what actually happened over the course of the century as carbon emissions rose. The models reproduce regional temperature trends over many decades, including warming trends since the 1950s and the cooling caused by large volcanic eruptions. We have increasingly detailed data from present-day measurements and from evidence about the past state of the climate from bubbles of air trapped in ice cores and from ocean sediments.

Could warming be due to natural forces? The answer is no. While natural forces such as volcanoes and variations in solar intensity can influence climate and have done so in the past, these natural variations cannot produce the currently observed patterns of climate change. In fact, many current natural variations seem to be in the direction of cooling—without them, warming due to human releases of greenhouse gases would be even greater. There has also been cooling due to some types of air pollution, which are decreasing due to control measures in many countries.

It may not seem plausible that small changes in the composition of the atmosphere could cause such big effects. Yet, the evidence does show that the climate system is sufficiently sensitive to atmospheric composition to produce the observed climate change, as shown by the response to other disturbances such as volcano eruptions (which pump gases high into the atmosphere). Resolving these issues eliminates some of the residual uncertainty that previously had clouded discussions of climate change, leaving little room for doubt that human-caused climate change is real and serious. Greenhouse gases may be a tiny fraction of the atmosphere, but this is hardly relevant: after all, insulation weighs almost nothing compared to the rest of the house, but it still plays an essential role in determining the temperature in the house.

For those of us who are not experts in climate science, there are limits to how well we can make independent judgments about the validity of the models now being used. Having done what we can to understand the basis for their judgments, at some point we must also give weight to the consensus among so many climate scientists regarding climate change projections. Given the broad convergence of all available models and observational evidence and the overwhelming agreement among the experts in projecting warming and its attendant effects, such as sea-level rise, current scientific findings are the best guide we can find.

Of course, complete scientific certainty is never possible but the current body of evidence comes close. With each IPCC report, scientists have ratchet up their level of confidence, with the latest report saying that the human contribution to warming is unequivocal. In any event, social policy can never be based on complete certainty. We make major governmental decisions based on social science evidence such as economic theories about inflation and unemployment that are subject to much less intensive scrutiny. Indeed, 95 percent is often used by experts in evidence law as a way of quantifying the concept of "proof beyond a reasonable doubt," and the evidence that humans are causing climate change has passed that threshold.

B. Where Do the Greenhouse Gases Come from?

To shape our response to climate change, we need to know about more than the role of greenhouse gases (GHGs) in driving the process. We also need to know where the greenhouse gases are coming from so we can design appropriate interventions.

Globally, about 23 percent of GHGs come from land use activities such as deforestation and agriculture.[3] The remainder comes from energy use, either through electricity generation, transportation, industry, or on-site use of fuels for heating or cooking. CO_2 is the dominant GHG. Besides CO_2, there are other greenhouse gases, most notably methane (often from animal waste or rice production), nitrous oxide (mostly from fertilizer use), and fluorinated gases (typically refrigerants).[4] These greenhouse gases can be quite potent, and the IPCC has created formulas to convert emissions of these gases into equivalent amounts of CO_2 (CO_2eq). Another contributor to climate change is black carbon (a particulate rather than a gas), often emitted by cookstoves and similar sources in developing countries. As of 2019, China accounted for 27 percent of global GHG emissions and the U.S. for another 13 percent. The EU and Japan accounted for another 11 percent. The rest was spread among many countries, including 5 and 7 percent respectively for Russia and India.

The figures are different, however, for cumulative historic emissions—as of 2015, the U.S. was at 20 percent and the EU nearly as high, while China was only one-half that amount cumulatively, even though China had higher total emissions in 2017 than the U.S. Given that these countries have different populations, the ranking of per capita emissions is different, with China still much lower than the United States. Whether past emissions are relevant to current day policy is a much-disputed issue that we discuss later in this chapter. The nature of the debate, too, may change, as cumulative

---

[3] P.R. Shukla et al., *Summary for Policymakers*, in IPCC. CLIMATE CHANGE AND LAND: AN IPCC SPECIAL REPORT ON CLIMATE CHANGE, DESERTIFICATION, LAND DEGRADATION, SUSTAINABLE LAND MANAGEMENT, FOOD SECURITY, AND GREENHOUSE GAS FLUXES IN TERRESTRIAL ECOSYSTEMS 8 (2019).

[4] EPA's website gives basic information on emissions, *see* http://epa.gov/climatechange/ghgemissions.

emissions from developing countries rapidly catch up with those from already developed countries.

There is little doubt that developed countries like the United States—and nowadays, emerging economies such as China—have contributed the most to global emissions. Those who may suffer the greatest damages are likely to be members of poor populations, sometimes within these very countries but often elsewhere in the world. Another group, which will suffer from climate change but lacks the power to influence present-day emissions, consists of members of future generations.

Within the U.S., the two biggest sources of emissions are transportation and electricity production. Electricity production generates a quarter of U.S. GHG emissions. Sixty percent of American electricity comes from burning fossil fuels, mostly coal and natural gas. Transportation accounts for over a quarter of domestic emissions. The great majority of transportation fuels are petroleum based. Industry accounts for another quarter, and commercial and residential buildings for about an eighth (separate and additional to their electricity use).[5] Energy-related U.S. CO_2 emissions peaked in 2000, remained roughly level until 2007, and since then have declined about 14%, thus falling back to the 1992 level.[6] Emissions per person have declined even faster, since there are a larger amount of people in the U.S. but smaller emissions. Annual emissions fluctuate due to changes in the economy, the price of fuel, and other factors such as the severity of the winter (and in 2020, COVID lockdowns).

Production of CO_2 by burning fossil fuels is clearly a critical part of the climate problem. In order to make major cuts in emissions, the transportation sector needs to move away from reliance on gasoline and oil diesel. At the same time, the electrical grid needs to be decarbonized. These steps are necessary but not sufficient to control greenhouse gases—there are other sources of CO_2 and other gases to worry about. But combustion of fossil fuels is the key. Changes in these emissions levels will be the major factor shaping our future climate. These changes will have relatively limited effect between now and 2050 because of "inertia" in the climate system, but much greater effects by the end of the century.

---

[5] These figures are from EPA's website, *see* http://epa.gov/climatechange/ghgemissions/sources.html.

[6] This figure is derived from the Energy Information Agency's website, https://www.eia.gov/environment/emissions/carbon/#:~:text=The%20decline%20in%20U.S.%20emissions,of%2015%25%20set%20in%202019.

C. What Does the Future Hold in Store?

We cannot change the past, only the present. So what we need to know is how climate change will look in the coming years. One of the key variables is within our control: the future trajectory of carbon emissions. The amount of future warming depends crucially on levels of GHG emissions. Even with very low emissions, warming would be one to almost two degrees (1.8–3.6°F) by the end of this century, while the increase could be 2.5–3.7°C with weaker, but still partially effective climate policies. The last time the earth experienced sustained global temperatures 2.5°C above preindustrial temperatures was three million years ago, well before our ancestors had discovered how to use fire. A century seems like an eternity, but there are children alive today who will live to see 2100. Because warmer temperatures melt glaciers around the world and cause the ocean's water to expand, these temperature changes correspond to a 0.5 to 0.75 meter sea-level rise above levels at the turn of this century according to the best current estimates. Even if we control emissions, the sea level will continue to rise for a long time as heat is mixed more evenly into the ocean. Thus, other generations will feel the impacts of our emissions far into the future.

Average temperatures won't rise evenly everywhere. Land will heat more than the oceans; continental areas more than coasts; and the arctic more than the tropics. But the real problem involves extremes rather than averages. Heat waves and severe droughts are already becoming more frequent, and so are floods. The subtropics which already tend to be arid, will get even less rain, while the far north will get more precipitation. We will discuss impacts in more detail in the chapter on climate adaptation.

Some concrete examples help to illuminate what is at stake if we do not control emissions. According to government projections, in a low emission scenario, by late in the century, the climate of New Hampshire will be similar to present day Maryland. With high emissions, New Hampshire's climate will resemble that of present day North Carolina.[7] With low emissions, Michigan turns into northern Arkansas by the end of the century, and with high emissions it will be more like northern Texas.[8] Similarly, under a high emissions scenario, temperatures will be higher than 90°F for more than half the year in South Florida[9] and South Texas.[10] The appearance of the countryside will change too. Under a medium

---

[7] U.S. GLOBAL CHANGE RESEARCH PROGRAM, GLOBAL CLIMATE CHANGE IMPACTS IN THE UNITED STATES 107 (2009), www.globalchange.gov/usimpacts.

[8] *Id.* at 117.

[9] *Id.* at 113.

[10] *Id.* at 34.

warming scenario, the maple and beech forests of the Northeast will be replaced by oak, pine and hickory, as will happen in much of the Great Lakes region. That vegetation is typical today of Tennessee.[11]

Although we know with high confidence that emissions will cause additional climate change, the extent of the warming is harder to predict. The problem is that there are many feedback loops in the earth's climate system, and many of them are not fully understood. One obvious feedback loop involves changes in the albedo, or the reflectivity, of the Earth's surface. Warmer weather in northern areas means melting snow and ice. Since the ground is darker than snow and ice, it will absorb more light and heat further, helping to melt other areas. On the other hand, if warming leads to more cloudy days, the clouds might reflect some of the light into space before it even reaches the ground. But to make things more complicated, nighttime clouds actually reflect radiation back to the earth, making the nights warmer. Climate modelers do their best to account for these feedbacks accurately, but the climate system is very complex and some of the feedbacks are understood better than others.

For these reasons, there is still uncertainty about the exact magnitude of climate change that a given level of greenhouse gas concentrations could cause. One useful measure of the relationship between GHG levels and the resulting warming is called climate sensitivity. It is defined as the amount of long-run temperature change (after all the feedback loops) caused by doubling the level of CO_2 in the atmosphere. According to the latest IPCC estimates, climate sensitivity is likely to be in the range 2.5°C to 4°C, with a best estimate of 3°C.

Science can provide information about the risks of climate change, but it cannot tell us how seriously we should take those risks or how vigorously we need to act. The science makes it clear that climate change will have serious disruptive effects for a long time to come. But what should we do about that?

Economists have developed some approaches for thinking about environmental problems that have become an important part of climate policy. Readers will differ in the extent to which they accept the validity of the economic approach. Even those who ultimately find it insufficient, however, may still find it useful as a rough way to categorize the tradeoffs involved in climate policy.

II. Climate Economics

U.S. law uses a variety of techniques to gauge the seriousness of risks and the effort necessary to control them. One of the tools used

---

[11] *Id.* at 18.

is economic analysis. Economics can help us understand the nature of the climate challenge, suggest possible instruments for controlling emissions, and hopefully shed light on the magnitude of the costs and benefits of controlling carbon emissions.

A. Climate Change as Externality

Economists think of climate change as a kind of externality. This simple concept should be part of everyone's intellectual toolkit. The key idea is that the market may not strike a proper balance between economic output and climate impacts because the costs of climate change do not fall directly on the emitters. Markets give companies incentives to adapt their behavior to current resources and consumer demands as signaled by prices. But this system only works to the extent that all the relevant factors are reflected in prices. As a result, "externalized" costs caused by climate change will not be taken into account by polluters. Without the incentive to reduce the amount of emissions, insufficient resources are devoted to this objective. Insofar as pollution costs are not borne by each emitter or its customers, some of the total welfare resulting from economic activity is redistributed away from the victims of climate change to other groups in the society. In a sense, the emitter is subsidized by others who bear the environmental costs of the emissions. Because of this subsidy, the emitter's conduct is not economically efficient: total social wealth (without regard for its distribution) is not maximized.

In economic terms, a firm that is able to externalize some of the costs relating to its production by emitting carbon is like a firm that receives a subsidy. For instance, if coal-fired power plants are not charged for their emissions, they are in effect being subsidized by the victims of climate change. In both cases, the firm is not paying the full cost of production, so it has an incentive to produce too much. Its production also shifts resources away from other firms. In the case of a firm receiving a cash subsidy, the resources are shifted from taxpayers directly, and from non-subsidized firms indirectly because the subsidized firm can charge an artificially cheap price. In the case of a firm that is allowed to emit carbon without paying for harm to victims, the effect is similar. Victims pay the direct price, while resource allocation is distorted because the emitter's prices do not include the full cost to society of its operations.

An economic activity may also confer external *benefits*. External benefits, like external costs, usually are not reflected in the producer's balance sheet. For example, basic research on clean technologies leads to knowledge that everyone can use. For that reason, it generally does not seem like a smart investment for businesses, since their competitors can use the knowledge as much as they do. The firms might be better off if they agreed to invest in basic research, but it does not make sense for any one of them to do so individually. Closely related to the concept of external benefits is that of "public goods," or commodities that cannot be supplied efficiently to one person without enabling others to enjoy them. Basic science is a public good in this sense. Consumption of such "public goods" is collective, because enjoyment by any one person does not diminish the enjoyment available to others and it is not practical to exclude anyone from the benefit. Because public goods involve positive externalities, they will be under-produced absent some kind of public intervention.

Because market signals are distorted when externalities are present, the government must intervene to limit external costs and facilitate production of external benefits and public goods. Much of what the government does falls into this category, whether it is protecting national security (a public good) or limiting water pollution (a negative externality). The following five basic approaches for public intervention are available, and all of them have been used or at least attempted in the context of climate change:

Liability. The common law doctrine of nuisance offers injured landowners the remedies of damages and injunctions against polluters. Making carbon emitters pay damages—or change their behavior—forces them to internalize the externalities they are creating. Efforts to sue carbon emitters on this basis will be discussed in Chapter 6. Liability issues are also relevant to international negotiations over what is called loss and damage—that is, the extent to which countries with high emissions should provide compensation for harms that cannot be avoided through adaptation measures.

Direct Regulation. The principal method of controlling pollution of all kinds has been to prohibit emissions beyond prescribed limits. Efforts by the Environmental Protection Agency (EPA) to regulate climate change in this way are also discussed in Chapter 6.

Subsidies. Another approach is for the government to subsidize private activities that reduce external costs or produce collective goods. In the area of climate change, examples include tax credits for renewable energy and financial support for energy research. During the aftermath of the 2008 Financial Crisis, the government also provided billions of dollars in direct subsidies to clean energy firms. More recently, the 2021 Infrastructure Act included $73 billion to update the nation's electricity grid so it can carry more renewable energy, $7.5 billion to construct electric vehicle charging stations, $17.5 billion for clean buses and ferries. These numbers were dwarfed, however, by the $369 billion in tax credits and direct funding for clean energy in the 2022 Inflation Reduction Act.

Charges. Besides the foregoing approaches, the government can require private actors to pay penalties or fees for activities that generate external costs or fail to provide external or public goods. Relating the amount of the charge to the estimated cost (to the producer) or benefit (to the public) can create a market incentive for the producer to alter his activities. A carbon tax is an example of an effluent charge. We will discuss carbon taxes in Chapter 4.

Pollution Markets. Rather than setting the level of charges itself, the government can use market mechanisms to put a price on carbon. Using hybrid systems consisting of regulation, effluent charges, and markets, the government can create tradable permits that firms can buy from and sell to each other. By limiting the total number of permits, the government "caps" emissions, but the distribution of pollution rights between various emitters is left to the market rather than being decided by the government. If the government auctions the allowances, emissions trading systems become very similar to emissions charges. Carbon trading is also discussed in Chapter 4.

These tools have different equity implications. Carbon taxes and auctioned carbon allowances place the economic burden on polluters, as does liability. Regulations also impose economic burdens on polluters, but in addition they ensure that each polluter will reduce emissions. This may result in improved air quality in some communities than taxes or trading would achieve. That emission reduction guarantee is absent from carbon taxes and trading systems, but in exchange they may allow greater emission reductions at the same cost. Under all these systems, many of the polluters' costs are likely to be passed on to consumers, sometimes placing a particular burden on the poor. Finally subsidies (whether through cash or tax credits) place the economic burden on taxpayers, which means that on average wealthier individuals bear more of the cost.

The discussion in this section indicates why climate change is harmful, why we cannot count on emitters to regulate themselves, and what tools are available for limiting emissions. But we also need to know the size of the externality. For policymaking purposes, we would like to know not only how much climate change to expect, but also what costs these changes will impose on society and what it would cost to ameliorate climate change. Unfortunately, our knowledge of these economic issues is limited.

B. Costs and Benefits of Climate Change Mitigation

In the case of climate change, one measure of the externality is the harm created by adding an additional ton of CO_2 to the atmosphere. This figure is called the social cost of carbon. "Social" because it includes damage to society as a whole, as opposed to a private cost that is experienced only by the emitter itself. Ideally, calculating costs in this way measures the harm caused by burning about a third of a ton of carbon (since one ton of CO_2 contains about that amount of pure carbon). Just as a rough gauge, a gallon of gas corresponds to twenty pounds of CO_2, and the average car produces just under five metric tons of CO_2 a year. So if the social cost of carbon is $60 today, the average car is producing $300 worth of climate harm per year.

Controlling CO_2 emissions may also result in controlling other forms of pollution. Combustion of fossil fuels is a leading cause of air pollution with coal being the worst offender and also the greatest culprit for CO_2 emissions. (Natural gas is a more efficient fuel, so the carbon released in order to generate a given amount of energy is lower than it is for coal.) Reductions in other pollutants produce significant health benefits. These "co-benefits" of reducing CO_2 are not included in the social cost of carbon, which only includes climate benefits. If we are considering the benefits of shifting from coal to natural gas, or from fossil fuels to wind and solar, however, we need to take into account not only the climate benefits but also the substantial co-benefits.

To figure out the social cost of carbon, first we need to estimate the amount of future warming using the climate models discussed in the last section, and then we need to figure out the cost associated with that amount of warming. That turns out to be a challenge. Ideally, we would like to be able to predict the future course of the economy in the absence of climate change and then redo the calculation taking into account climate impacts. But our ability to make even the first type of prediction is limited, and there is considerable uncertainty about the economic effects of climate change. This uncertainty is due in part to incomplete understanding of the physical impacts of climate change, as well as our inability to predict how successfully people will learn to avoid the resulting costs through adaptation and how much adaptation itself will cost. We will discuss this in more detail in the chapter on adaptation.

There are now many models that combine climate change predictions and economic analysis. These models differ along a number of dimensions, including: their focus on the energy sector or reliance on a broad macroeconomic analysis, the degree to which they analyze localized versus average global impacts, and their treatment of uncertainty. Model results differ correspondingly. The resulting estimates of the social cost of carbon are better than blind guesses but still a long way from hard numbers. There are similar difficulties in modeling the costs of mitigating and adapting to climate change. Hopefully, economists will be able to narrow the uncertainty, but at this point there is a wide range of disagreement.

Many of the individual elements of the economic impact analysis are the subject of serious debate. For instance, economists hotly dispute the net effect of climate change on agriculture, with some finding an overall positive effect on U.S. agriculture (but with very large regional variations), and others finding substantial negative effects. If we do not even know whether important elements of the economic impact count as benefits or costs, predicting overall impact (taking into account all of the feedback loops of the economy) is obviously going to be difficult.

Modeling the systemic economic impact of climate change as well as the costs of adaptation and mitigation entails tremendous challenges, particularly if the projection extends more than a few years. To begin with, the economic model must build on the outputs of climate models, which are themselves uncertain. Then there is the difficulty of forecasting the future trajectory of the economy over future decades. This clearly cannot be done in detail—for example, no forecaster 50 years ago would have predicted the explosive growth of personal computers, let alone the Internet.

Even more rudimentary forecasts rely heavily on the assumption that the future will on average be much like the recent past—for example, that technological progress will continue around its current pace and that some unforeseen catastrophe will not cause an economic crash. Even predictions for specific economic sectors are difficult. Past experience with models that project energy use do not lend much confidence to these predictions: the projections have generally been too high, by as much as a factor of two. As an example, the advent of hydraulic fracturing and its dramatic effect on the price of natural gas were not foreseen. Nor were spectacular decreases in the costs of renewable energy and battery storage. Projecting adaptation is made more difficult by institutional barriers that may prevent deployment of optimal adaptation strategies. For instance, political disputes may prevent timely construction of flood barriers, so that major flood damage occurs before they are complete. The uncertainties go both ways: to the extent that climate change scenarios are based on projections of future emissions, they implicitly make assumptions about future political and economic developments.

Overall, economic modeling is still at a primitive stage compared with climate modeling. This is largely unavoidable—our knowledge of human behavior is far less developed than our knowledge of physical processes, and so our corresponding ability to predict human behavior is less advanced than our ability to predict physical processes.

Even if we knew all the economic costs, including those created by climate impacts and those relating to mitigation, we would still face some serious challenges. The first is that not all of the impacts are financial and, thus, there are valuation problems; the second is that we need to account for the fact that climate impacts will be spread out over decades and even centuries to come; and the third is that we need to factor in the possibility of low-probability but potentially catastrophic outcomes (such as global changes over 6°C or severe local effects, such as loss of monsoons in India).

Valuation Problems. It's easy to see how we can put a monetary value on property damage or lost economic production. But how do we put a price on health risks or ecological harm? Without worrying right now about the technicalities, we could assign monetary values to different levels of risk by looking at how much consumers are willing to pay for safer products, or how much income workers are willing to give up for safer jobs, or how much travel time people are willing to sacrifice for the safety benefits of driving more slowly. All of these would be different ways of determining the market value of safety.

If people demand $1,000 in return for being exposed to a one in a thousand risk of death, it's conventional to say that the "value of life" is $1 million. This is a bit misleading, since they probably wouldn't be willing to die for that amount of money! To express this distinction, economists often speak of the value of a statistical (as opposed to individual) life. To assign a value of $1 million per statistical life is the same as saying that people would demand $1 million in return for running a one-in-a-million chance of death. EPA currently estimates that the value of a statistical life is about $10 million, but this is only an estimate. Obviously, there are legitimate ethical arguments against putting a price tag on life at all, but we're deferring consideration of those arguments for the moment.

Climate change will also have dramatic impacts on ecological systems. These ecological impacts are hard to predict and hard to translate into monetary equivalents. For instance, what price should we put on the extinction of polar bears? Economists have developed some techniques to try to deal with this issue, but it is fair to say that they remain controversial.

The Time Factor. A ton of CO_2 emitted today will stay in the atmosphere for two to three centuries, continuing to cause climate impacts. Surface temperatures will remain high even after we eliminate emissions for an even longer time because of the excess heat stored in the ocean (which will also lead to continuing sea level rise). Conversely, a dollar invested in reducing CO_2 emissions today will provide benefits over the same period of time. In assessing the value of this investment, we need to take into account the long timespan involved.

Economists use a technique called discounting for this purpose. This can be a confusing concept, but it is based on a simple observation about human nature. The basic idea is that, for a variety of reasons, people have a preference for receiving benefits earlier rather than later, and a corresponding preference for postponing costs. For example, suppose you are given a choice between getting some money today and $100 a year from now. Let's adjust for inflation so we don't need to take that into account. Even ignoring inflation, most people will be willing to take less than a $100 dollars today instead of waiting a year. This can be due to several factors—one of them is simple impatience, another is that you might get a better return by investing the money elsewhere, and yet another is that you might expect to have more money next year anyway, so you need the money more badly today. Let's say that you would take $90 today, so that $90 today is viewed as being equivalent to $100 next year.

That means you would only make an investment of $90 this year if the payout is at least $100 in one year's time, an 11 percent increase. How much would you want today in exchange for $100 in *two* years? Since you're demanding an 11 percent rate of return we have to discount $90 by 11 percent to take account of the extra year of delay. We know that $100 in year 2 equals $90 in year 1, and in turn that equals about $82 dollars today (111 percent of $82 is just over $90). So, $100 dollars in two years equals only $82 today. One way of expressing this is to say that the "present value" of the future payment of $100 in two years is only $82 dollars today. This kind of calculation is familiar in the business world when deciding whether an investment today is worthwhile in terms of future payoff.

Don't worry if your eyes glazed over when you were reading the numbers. The key point is that the value of a future dollar falls over time in accordance with the discount rate. Over a long period of time—the kind of time period involved with climate change—the changes are really dramatic. For instance, at a 10 percent discount rate, $100,000 a century from now equates to about $7.30 today! Thus, using the same discount rate, if we knew that a ton of carbon emitted today would cause $1,000 worth of damage a century from now, we would only be willing to pay seven cents to avoid this harm. Another way of seeing the effect of discounting is that if a ton of carbon caused $1,000 in damage every single year forever, we would only be willing to pay $10,000 to avoid the harm with a 10 percent discount rate. That's equal to the damages for only the first ten years, as compared to the huge cumulative amount of damage after that time period.

Given the dramatic impact of discounting on long-range issues like climate change, it's not surprising that there is considerable controversy about whether this technique is justifiable. As one economist has said, discounting "forces us to say that what we might otherwise conceptualize as monumental events 'do not much matter' when they occur in future centuries or millennia."[12] Even if we assume that discounting is appropriate when dealing with the effect of policies on current generations, its application to future generations seems to raise graver ethical issues. After all, their value as human beings is equal to that of people today, yet discounting systematically downgrades their interests, sometimes to the point of insignificance.

One defense of discounting is that funds spent to avoid climate change could be spent on other purposes, which might be more beneficial to future generations. There is widespread agreement, even among critics of discounting, that these "opportunity" costs deserve consideration. Thus, we might want to engage in discounting in the interest of future generations in order to maximize the benefits to them of present-day investments. Market rates represent the opportunity cost of investment, so this argument suggests that we should avoid climate mitigation projects unless they offer equal returns. One problem with this argument is that climate change might have catastrophic effects on members of later generations that cannot be offset by increased savings.

Another argument for discounting is that, given economic growth, future generations are likely to be much wealthier than current generations, even taking into account climate change. Why should present generations sacrifice to make future generations even richer—isn't this in effect transferring money from the (relatively) poor to the (relatively) rich? A counterargument is that, although technological progress has driven economic growth for an extended period of time, we have no assurance that this mechanism will continue to operate. Even today, although technology is a constant, individual countries have much different records of growth, which suggests that other factors are also crucial. Given the limits on how well we understand growth, we may not have any real assurance that present growth rates will continue.

If discounting is indeed a valid approach, other, more technical difficulties must be confronted. As it turns out, the discount rate is very important; small differences can make a big difference in the outcome. You can see this in the following table, which includes numbers that the U.S. government has used at various points:

---

[12] Martin L. Weitzman, *Why the Far-Distant Future Should be Discounted at Its Lowest Possible Rate*, 36 J. ENV. ECON. & MANAGEMENT 201, 201 (1998).

Effect of Discount Rate on Present Value

	Present value of $6 million received in 20 years	Comments
10%	$894,000	The rate used 1972–1992 by government.
7%	$1,560,000	Newer rate used after 1992.
3%	$3,324,000	Optional rate allowed after 2003.

As you can see from the table, cutting the discount rate from seven to three percent more than triples the present value of the eventual benefit, from $894,000 to $3,324,000. Cutting the rate another percent would bring the present values up to $4,000,000. Over longer periods of time, the differences are even more dramatic. For instance, over forty years, raising the discount rate from three percent to seven percent reduces the present value almost 80 percent, from $1,842,000 to $405,600. Over eighty years, the difference is $559,494 versus $26,934, an almost twenty-fold difference.

The bottom line is that discounting makes a big difference in assessing costs and benefits over long periods of time, and that difference is quite sensitive to changes in rates. Over multiple-decadal time scales, a minor shift in the discount rate can dramatically impact the analysis of whether additional precautions are warranted.

The trouble is that economists don't agree on the right discount rate. For this reason, in trying to calculate the social cost of carbon, the U.S. government has used multiple discount rates. The following table provides the results from a 2020 update of the figures generated during the Obama Administration.[13]

[13] Interagency Working Group on Social Cost of Greenhouse Gases, *Support Document: Social Cost of Carbon, Methane, and Nitrous Oxide—Interim Estimates under Executive Order 13990* 5 (Feb. 2021).

Table ES-1: Social Cost of CO_2, 2020–2050 (in 2020 dollars per metric ton of CO_2)

Emissions Year	Discount Rate and Statistic			
	5% Average	3% Average	2.5% Average	3% 95th Percentile
2020	14	51	76	152
2025	17	56	83	169
2030	19	62	89	187
2035	22	67	96	206
2040	25	73	103	225
2045	28	79	110	242
2050	32	85	116	260

Notice that the numbers are quite different horizontally, depending on the discount rate. But the number in each column rises over the years. That's because as time goes on, we get closer to the times when the costs of climate change mount up, so we're discounting those heavy damages over a shorter time period. In short, as time goes by, we would find it worthwhile to spend more and more to counter climate change because we're dealing with harms that are increasingly imminent and costly.

The final column combines a 3% discount rate with a safety margin to account for the risk that we will be unlucky. It assumes that the harm will be at the top 95th percentile of all the different estimates, instead of in the middle. This brings us to the key economic factor, the extent to which we should pay more to eliminate climate change as insurance against the risk of "unlucky" outcomes due to high feedback.

The "Insurance" Factor. William Nordhaus, who won a Nobel prize for pioneering economic models of climate change, explains some of the serious risks associated with climate change:

> There are continuing open questions about the future of the huge ice sheets of Greenland and West Antarctica; the impact of aerosols on global and regional climates; the risks in thawing of vast deposits of the frozen methane and permafrost; changes in the circulation patterns of the North Atlantic; and the impacts of ocean carbonization and acidification. . . .

> [W]e might think of the large-scale risks as a kind of planetary roulette. Every year that we inject more CO2 into the atmosphere, we spin the planetary roulette wheel. . . .
>
> A sensible strategy would suggest an insurance premium to avoid the roulette wheel in the Climate Casino. . . . We need to incorporate a risk premium not only to cover the known uncertainties such as those involving climate sensitivity and health risks but also the zero and double zero uncertainties such as tipping points, including ones that are not yet discovered.[14]

The difficulty, as Nordhaus admits, is trying to figure out the extent of the premium. Another book by two leading climate economists argues that the downside risks are so great that "the appropriate price on carbon is one that will make us comfortable enough to know that we will *never* get to anything close to 6°C (11°F) and certain eventual catastrophe."[15] Although they admit that "never" is a bit of an overstatement—reducing risks to zero is impractical—they clearly think it should be kept as low as feasibly possible. Not all economists would agree with that view. However, there seems to be a growing consensus that the possibility of catastrophic outcomes should play a major role in determining the price on carbon. If you look back at the social cost of carbon table given above, you'll see that the even without taking tipping points into account, the high-risk "95th percentile" figure is triple the average estimate using the same discount rate.

The extent of disagreement among economists raises the question of why we should even try to quantify the cost of climate change. The best argument for doing so is that we need some basis for deciding how much society should spend to control emissions. Even if the number is uncertain, the alternative to using some number for climate change in cost-benefit analysis is essentially to set the number to zero, and scientists consider that a very implausible estimate. Although the exact magnitude of the harm is unknown, due to both scientific and economic uncertainty, estimates by economists seem as likely to be too low as too high. Thus, they may represent a reasonable middle-of-the-road basis for policy decisions.

One lesson of the economists' efforts may simply be that even a fairly dollar-and-cents approach supports robust action on climate change. On the whole, efforts to set the social price of carbon are rather cautious. They use discounting to offset the long-time horizon

---

[14] WILLIAM NORDHAUS, THE CLIMATE CASINO: RISK, UNCERTAINTY, AND ECONOMICS FOR A WARMING WORLD 142–43 (2013).

[15] GERNOT WAGNER & MARTIN L. WEITZMAN, CLIMATE SHOCK: THE ECONOMIC CONSEQUENCES OF A HOTTER PLANET 78 (2015).

for climate change, and so far, the risk of catastrophic outcomes does not figure heavily in the models. They generally also leave out any concerns we might have about the justice of certain countries causing climate change that disproportionately will harm other, low GHG-emitting countries. This is a factor that looms large in international negotiations, and we turn to it now in order to complete our overview of the basics of climate policy.

C. Government Estimates of the Social Cost of Carbon

The Obama Administration made a concerted effort to estimate the social cost of carbon, resulting in the table shown earlier in this chapter. Administration officials were well aware of the economic uncertainties, and it is worth explaining how they attempted to deal with the problem. Obama's Interagency Working Group (IWG) began by selecting the three most widely cited models used by economists. Each of those models combine a climate change model with an economic model, measuring the amount of damage from carbon emissions under various scenarios regarding economic growth and carbon prices. One of the models tends to produce low estimates of climate harm; one is much higher; and the third is in the middle.

These models had to be tweaked to make them comparable. The IWG decided to focus on the *global* impacts of carbon, partly for policy reasons, but also because the vast majority of economics literature does so. The IWG then used various discount rates, a crucial factor in calculating the social cost of carbon, to provide a range of estimates. In order to make these models compatible, the IWG had to do some very sophisticated tweaking and recalibration. They then ran the models a large number of times to get the full range of outcomes, in order to take account of the range of uncertainties about how CO_2 levels might evolve and how the climate may respond. To produce the final figures, the model outputs were averaged to predict the amount of economic harm in each future year. Then, the IWG applied several different discount rates because of disagreements about which rate is correct, and the IWG also did the same calculations for the upper 5 percent of the range of harms, to take into account the risk of catastrophic outcomes. All of the resulting figures were reported to the public, as shown in the table earlier in the preceding section.

These estimates were upheld by the Seventh Circuit in *Zero Zone, Inc. v. DOE.*[16] Challengers to energy efficiency standards for commercial refrigeration attacked the use of the social cost of carbon in setting the standard. The court concluded that "DOE's

---

16 832 F.3d 654 (7th Cir. 2016).

determination of SCC was neither arbitrary nor capricious."[17] In particular, the court rejected the argument that the social cost of carbon could include only harm within the United States: "According to DOE, national energy conservation has global effects, and, therefore, those global effects are an appropriate consideration when looking at a national policy. Further, [the challengers] point to no global costs that should have been considered alongside these benefits. Therefore, DOE acted reasonably when it compared global benefits to national costs."[18]

Despite this judicial endorsement, the Obama Administration's estimate of the social cost of carbon remained unpopular with Republicans. Two months after taking office, President Trump issued an executive order rescinding the Obama-era estimates.[19] Instead, Trump directed agencies to make their own determinations of the social cost of carbon, which could involve using a higher discount rate and considering only costs within the United States (which involves technical difficulties because of the way that current climate models are designed). The Trump Administration's estimate was 90% lower than the Obama Administration. The Biden Administration went back to the global approach.

These shifts in methodology have given rise to litigation, with no definitive resolution to date. In 2020, a California district court invalidated a Trump Administration estimate of the social cost of methane. The court said that the agency had failed to take into account the indirect effects of the foreign impacts of climate change on the United States and had not explained the reasons for its rejection of the consensus among economists for including those effects.[20] In 2022, though, a Louisiana district court enjoined the Biden Administration from developing new estimates of the social cost of carbon on a variety of grounds, including that the estimates would include non-U.S. impacts of climate change.[21] However, the Fifth Circuit stayed the injunction because the plaintiffs had failed to establish standing. Another district court found that similar claims were not ripe for review, because no regulatory action as yet had been based on revised estimates.[22] No doubt the issue will arise again once the Biden Administration's estimate of the social cost of carbon is actually put into use.

---

[17] *Id.* at 678.

[18] *Id.* at 679.

[19] Presidential Executive Order on Promoting Energy Independence and Economic Growth (Match 28, 2017). The provisions relating to the social cost of carbon are in section 5 of the order.

[20] California v. Bernhardt, 472 F. Supp. 3d 573 (N.D. Cal. 2020).

[21] Louisiana v. Biden, 2022 WL 438313, at *16 (W.D. La. Feb. 11, 2022).

[22] Missouri v. Biden, 558 F. Supp. 3d 754 (E.D. Mo. 2021).

D. Climate Equity

Emitters of GHGs will cause serious harm, especially in the poorest countries, which often have very low emissions themselves. Poor countries are vulnerable partly because they often happen to be located in places where they are exposed to major climate impacts. They may be less able to protect themselves from those impacts due to weak social, political, and economic infrastructure. For example, they may not be able to afford expensive sea walls or the expense of rebuilding after extreme weather events. Even those who discount the relevance of such ethical considerations concede that "emissions in some countries have imposed serious risks on others, that the United States and China are expected to remain the world's leading contributors, and that some nations, including those in Africa, face serious risks even though their own emissions are trivial."[23]

In this vein, there are several possible arguments for why some countries should be expected to shoulder more of the burden of climate mitigation than others:

1. The "tort" argument: Through their emissions, countries with high emissions have caused harm to others. Whether or not this is technically a basis for tort liability, they should be held responsible for the resulting harm.

2. The "distributional" argument: By and large, the high emitting countries are also the richer countries. They should bear a greater responsibility than poorer countries that can less easily absorb the costs of carbon reductions. Economists have devised a technique called equity weighting for taking these distributional effects into account. Using that technique would greatly increase the social cost of carbon and thus require much greater efforts by developed countries to cut emissions.

3. The "common resource" argument: The planet's capacity to handle increased carbon is limited, and that limited capacity should be shared equally by all humans. In effect, humanity has a limited "budget" of carbon emissions, and that budget should be shared equally. As a practical matter, adopting this approach would mean that many developing countries could increase their per capita carbon emissions from their current very low levels, while developed countries would need to make massive cuts to get down to the global average.

---

[23] ERIC POSNER & DAVID WEISBACH, CLIMATE CHANGE JUSTICE 101 (2010).

Each of these arguments has given rise to extensive debate, which would take us far beyond the confines of this book. We will merely sketch some contours of the debate among legal scholars over these arguments.

For instance, the tort argument may rest on an assumption that each individual rich country is to blame for the harm caused by their emissions. But a counterargument is that a unilateral reduction in emissions by any individual country would have had no benefit. This counterargument is sometimes phrased as an assertion that the amount of emissions from any single country is too small to matter, compared to the entire amount in the atmosphere. That assertion seems to be false. In terms of the factual issue, climate models show that any increase in emissions, regardless of the existing level, causes incremental harm. Unless restrictions on emissions in rich countries would have caused other countries' emissions to increase enough in response to fully offset the reductions ("carbon leakage"), any reduction in emissions would decrease or at least delay damage from climate change even if a rich country acted alone. Thus, the argument that reducing emissions would have been pointless has to rest on some strong assumptions about carbon leakage. The empirical evidence does support the existence of leakage, but not at the level necessary to wipe out reductions in countries that are controlling emissions.

Furthermore, proponents of the tort analogy may argue, even if no single country could unilaterally temper climate change, it does not necessarily follow that each country is free from responsibility for the harm caused by its emissions. Tort law has long rejected the argument that, when harm proceeds from multiple sources, individuals can avoid responsibility by showing that it would have made no difference if they alone had acted properly. For instance, if two actors independently start fires negligently, neither one can escape liability by arguing that the other fire would have caused the same harm anyway.[24] The general tort rule is that "at least where both causes involve comparable blameworthiness, both actors are liable, even though the conduct of either one was not a sine qua non of the injury because of the conduct of the other."[25] Thus, even if action by a single country would have had little effect assuming other nations continued unrestrained emissions, nations are not necessarily free from responsibility for their excessive levels of emissions.

---

[24] *See* KENNETH S. ABRAHAM, THE FORMS AND FUNCTIONS OF TORT LAW 111 (3d ed. 2007) (stating that this is the "universal" outcome and that it would be "absurd" to relieve either negligent party of liability).

[25] Boeing Co. v. Cascade Corp., 207 F.3d 1177, 1183 (9th Cir. 2000).

Still, even if rich country reductions would have reduced or delayed climate harm, one could argue that it is unfair to impose costs that will ultimately be placed on the individual citizens of those countries, since they may not have been individually at fault for failing to reduce emissions. Moreover, at least some of those residents may in fact have reduced their own emissions, and others may have had low emissions because they are poor (and on average poor people generate fewer emissions than rich people.) In addition, one might contend, even poorer countries generally have at least some wealthy residents with high responsibility for emissions. Hence, imposing responsibility at the national level is arguably unfair to individual citizens of those countries. Naturally, advocates of the tort analogy have responses of their own to this counterargument.

The distributional argument has likewise been challenged. Opponents argue that if we want to help poor countries, climate change mitigation is a poorly designed form of foreign aid. Why not just give poorer countries financial assistance? Again, there are counterarguments. Simply transferring large amounts of cash may not be a good way to improve the welfare of people in poor countries, and in any event is unlikely to happen. Opponents of the distributional argument have responses to this point, and so the debate goes on.

The "carbon budget" argument is somewhat newer and, thus, the dimensions of the argument and counterarguments are still emerging. However, there are two major counterarguments. A strong practical criticism is that developed countries like the U.S. would never agree to such draconian cuts. A more theoretical criticism is that equality ought to be at an individual level: higher income individuals, no matter where they are located, should have the same responsibility to contribute to addressing climate change. Given that some emerging economies such as China have very unequal distributions of income and emissions levels, this argument might result in treating some Chinese citizens on a par with citizens of developed countries, while others would be treated as on a par with much less developed countries.

We have only skimmed the surface of the debate over climate equity. Economists are mostly concerned with problems of efficiency—how much should global emissions be cut, and what is the cheapest way of making those cuts? Economics has much less to say about who should ultimately bear the costs, for instance whether developed countries should help pay for reductions in the emissions of developing countries. Yet considerations of fairness are crucial in terms of reaching a consensus on the action to be taken in combatting climate change.

E. Climate Change and the Problem of Collective Action

Pollution control may be an apt example of what game theorists call a prisoner's dilemma. This scenario gets its name from the following illustrative story. Consider two prisoners, charged with the same crime and held in different rooms. The prosecutor gives each of them the following information:

> a. If neither of you confesses, I will charge you both with a lesser crime that I can easily prove, resulting in two-year sentences for each.
>
> b. If you both confess, you will each get a three-year sentence.
>
> c. If only one of you confesses, that person will get a one-year sentence and the other will get a four-year sentence.

Because the prisoners cannot communicate (and might not be able to trust each other even if they could), each one must decide what to do without knowing what the other will do. Each prisoner reasons as follows: "if the other person confesses, I should do the same. That way, I would get a three-year sentence instead of a four-year sentence. If the other person doesn't confess, I'm still better off if I confess, because that would give me a one-year sentence rather than a two-year sentence. Thus, no matter what, I will be better off if I confess." Following the same reasoning, the other prisoner also confesses, and they both get three-year sentences. This is the "rational" outcome. But notice that if the prisoners could somehow count on each other, neither would confess, and they would get only two-year sentences. So the individually rational set of actions leads to an inferior outcome in terms of their group welfare. The scenario itself is contrived and unrealistic. The "prisoner's dilemma" tag for this phenomenon has stuck, however. Perhaps one reason is that it points up how we are all prisoners of our inability to cooperate for the common good.

What does this have to do with climate change? Consider the situation of Freedonia, a hypothetical average country. If the rest of the world fails to address the greenhouse effect, Freedonia can do little on its own, and much of the benefit of its effort would go to other countries that aren't even trying. Therefore, it shouldn't bother. If everyone else *does* take action to control the greenhouse effect, Freedonia can contribute only slight additional help but will have to spend a lot of money to do so. So, if everyone else "does the right thing," Freedonia should take a "free ride" on their efforts rather than wasting its own resources to minimal effect. Thus, no matter what the rest of the world does, Freedonia is better off to do nothing.

Reasoning the same way, every country in the world decides to take no action. This "climate dilemma" seems much like the prisoner's dilemma discussed above.

Yet, all countries might be better off if they could somehow make an enforceable deal to cooperate. But such deals are difficult to negotiate and enforce on the international level, because the prisoner's dilemma takes hold again: each individual country has an incentive to sit out the negotiations and let everyone else make a deal; once a deal is made, each country has an incentive to breach the agreement and allow others to bear the costs of carrying it out.

The upshot is that, in the absence of some mechanism for making and enforcing cooperative agreements, each nation may find inaction to be the only sensible individual choice, though everyone also knows that collective inaction will only lead to disaster. Within individual nations, the central government can lead the way out of prisoner's dilemmas by forcing a cooperative solution. But climate change is a global problem, requiring a global solution.

On its face, climate change seems to present a nearly insuperable problem of collective action. However, as we will see in the next chapter, there has been actual progress in international cooperation. Perhaps even more surprisingly, individual nations and even sub-national units have taken action to address emissions without waiting for an international agreement. Is it worthwhile to take action on climate change even without waiting for a global agreement?

There are several reasons why climate change may not quite fit the model of the prisoner's dilemma. The first is that unilateral action by large emitters can be worthwhile for them even if not every country is on board, so long as three conditions are met: (1) the coalition must account for a large enough share of global emissions that its collective reductions will have a measurable impact on emissions, (2) the potential climate damages must be high enough that the resulting benefit to coalition members exceeds their emission control costs, and (3) the coalition's efforts to reduce emissions can't result in an offsetting increase in emissions by non-coalition members. In practical terms, this means that a limited number of key nations such as the U.S., the countries within the EU, and China, could form an effective coalition if they can prevent "carbon leakage" to other countries. Furthermore, unlike the prisoners in the hypothetical, countries have the opportunity to cooperate over time and gain trust in each other. Finally, unlike the hypothetical prisoners, countries can use a variety of other levers to augment cooperation, such as economic sanctions. So the collective action problem remains difficult, but it is not really as bad as a true prisoner's dilemma.

In the next chapter, we turn to efforts to overcome the collective action problem through international negotiations. As discussed in this chapter, the advancing science makes it clear that climate change will cause serious harm, even though the exact amount is uncertain. Economic analysis confirms that at least some substantial response is warranted. Just how much of a response is warranted economically depends heavily on two key factors—how much weight we give to climate impacts after the next fifty or sixty years, and how much we are willing to pay as insurance against tipping points or catastrophic outcomes. The issues are not merely economic, but also relate to ethical views about duties to our descendants and to poorer nations. However we set our objective, many barriers will need to be surmounted in order to achieve it. The next five chapters are devoted to pathways for achieving mitigation goals.

Further Readings

Cinnamon Carlarne, *Climate Courage: The Remaking of Environmental Law*, 41 STAN. ENV. L.J. 125 (2022).

Daniel Cole, *The Stern Review and Its Critics: Implications for the Theory and Practice of Benefit-Cost Analysis*, 48 NAT. RES. J. 53 (2008).

Daniel A. Farber, *Coping with Uncertainty: Cost-Benefit Analysis, The Precautionary Principle, and Climate Change*, 90 WASH. L. REV. 1659 (2015).

KATHARINE HAYHOE, SAVING US: A CLIMATE SCIENTIST'S CASE FOR HOPE AND HEALING IN A DIVIDED WORLD (2021).

INTERGOVERNMENTAL PANEL ON CLIMATE CHANGE, 2014 SYNTHESIS REPORT: SUMMARY FOR POLICYMAKERS (2014).

WILLIAM NORDHAUS, THE CLIMATE CASINO: RISK, UNCERTAINTY, AND ECONOMICS FOR A WARMING WORLD (2013).

ERIC POSNER AND DAVID WEISBACH, CLIMATE CHANGE JUSTICE (2010).

J.B. Ruhl & Robin Kundis Craig, *4°C*, 106 MINN. L. REV. 191, (2021).

NICHOLAS STERN, THE ECONOMICS OF CLIMATE CHANGE: THE STERN REVIEW (2007).

U.S. GLOBAL CHANGE RESEARCH PROGRAM, GLOBAL CLIMATE CHANGE IMPACTS IN THE UNITED STATES (2018).

GERNOT WAGNER AND MARTIN L. WEITZMAN, CLIMATE SHOCK: THE ECONOMIC CONSEQUENCES OF A HOTTER PLANET (2015).

Chapter 3

THE INTERNATIONAL CLIMATE CHANGE REGIME

The international climate change regime revolves around three primary instruments of international law—the United Nations Framework Convention on Climate Change (UNFCCC),[1] the Kyoto Protocol to the UNFCCC,[2] and the 2015 Paris Agreement.[3] This chapter explores the history and evolution of the international climate regime.[4] It proceeds in four parts. First, it examines the emergence and evolution of the regime through past and present international climate negotiations. Second, it reviews the general principles around which the regime revolves. Third, it introduces the Kyoto Protocol's market-based mitigation mechanisms and discusses how the Paris Agreement envisions the future of cooperative mechanisms, including a discussion of the ongoing development of the Paris "rulebook" and the Glasgow Climate Pact.[5] Fourth, it concludes with a discussion of the future of the international climate change regime.

As background to this chapter, it is important to keep in mind the sharp difference between international and domestic law. In a country like the United States, new laws can be made by a majority vote of legislators. There is no equivalent to the U.S. Congress in the international sphere, and as a result, treaties are only binding on nations that ratify them. Thus, in the case of climate change, an effective solution requires voluntary cooperation on a global scale. Obtaining effective climate policy even in a single country is difficult enough. It is even more difficult to obtain agreement from countries across the world with very different economies, governance systems, and political dynamics. It should not be a surprise that process has been protracted.

---

1 May 9, 1992, 1771 U.N.T.S. 107, *reprinted in* 31 I.L.M. 849 (1992) [hereinafter UNFCCC].

2 Kyoto Protocol to the United Nations Framework Convention on Climate Change, Dec. 10, 1997, UN Doc. FCCC/CP/1997/7/Add.2, Dec. 10, 1997, *reprinted in* 37 I.L.M. 22 (1998) [hereinafter Kyoto Protocol]. The Kyoto Protocol opened for signature March 16, 1998, and entered into force February 16, 2005.

3 Paris Agreement, Preamble, Dec. 12, 2015, U.N. Doc. FCCC/CP/2015/L.9/Rev.1 (entered into force Nov. 4, 2016).

4 The term 'Regime' is used here to refer to the UFCCC, the Kyoto Protocol, and associated instruments and decisions.

5 Glasgow Climate Pact, UNFCCC Decision -/CMA.3 (Nov. 13, 2021) [https://perma.cc/2GJV-3GCQ].

I. Brief History of the International Climate Change Regime

A. The United Nations Framework Convention on Climate Change

In 1990, the United Nations General Assembly (UNGA) initiated the development of the field of international climate change law with Resolution 45/212. This Resolution launched international negotiations for a multilateral climate treaty.[6] Within two years, these negotiations culminated in the adoption of the UNFCCC. The UNFCCC is a Framework multilateral treaty open to all nations and to the European Union as a regional association. The strategy behind use of a framework convention is to get agreement first on an objective and some basic principles and establishing a negotiating structure for converting those principles into more concrete obligations.

The objective of the Convention is to achieve "stabilization of greenhouse gas concentrations in the atmosphere at a level that would prevent dangerous anthropogenic interference with the climate system" and to do so "within a time-frame sufficient to allow ecosystems to adapt naturally to climate change, to ensure that food production is not threatened and to enable economic development to proceed in a sustainable manner."[7]

Beyond defining this overarching objective, the UNFCCC enumerates a set of fundamental principles to guide efforts to address climate change; establishes a series of common commitments, as well as a series of commitments specific to countries with higher levels of economic development; and creates a set of rules and institutions designed to support efforts to implement the Convention and meet the goal of stabilizing global greenhouse gas emissions. Since coming into force in 1994,[8] the UNFCCC, through the decisions of the Parties to the Convention, has been the primary driver of international climate change law, including the development of implementation and compliance tools, financing instruments, and basic norms and principles.

The important role that the UNFCCC has played is not surprising given that participation in the Convention is "near-

---

[6] Protection of Global Climate for Present and Future Generations of Mankind, G.A. Res. 45/212, U.N. Doc. A/RES/45/212 (Dec. 21, 1990).

[7] UNFCCC, *supra* note 1, art 2.

[8] UNFCCC, *supra* note 1. Opened for signature May 9, 1992, and entered into force Mar. 21, 1994.

universal," with 197 parties having ratified the treaty.[9] This means that 197 states and one regional economic organization (i.e., the European Union)—representing all UN Member States—are parties to the treaty. (It is worth noting that although the United States did not become a party to the Kyoto Protocol, as discussed below, it did sign and ratify the UNFCCC under the leadership of President George H.W. Bush.) By becoming parties to the Convention, each of these entities has formally recognized that "human activities have been substantially increasing the atmospheric concentrations of greenhouse gases, that these increases enhance the natural greenhouse effect, and that this will result on average in an additional warming of the Earth's surface and atmosphere and may adversely affect natural ecosystems and humankind."[10] Based upon recognition of the problem, each party to the treaty has agreed, at a minimum, to begin to take basic steps to quantify and report on national greenhouse gas emissions and to develop national and regional programs "containing measures to mitigate climate change."[11]

Although the UNFCCC is the central international climate change legal instrument, because it is a framework convention it focuses on establishing overarching objectives and norms rather than establishing detailed implementation measures and state- or category-specific legal obligations. As a result, upon adoption of the UNFCCC, the parties to the treaty began negotiations at the first annual Conference of the Parties (COP) for a supplemental legal instrument that would provide greater precision as to how to meet the objectives of the treaty.

B. The Kyoto Protocol

In 1997, three years after the UNFCCC entered into force, the COP adopted the Kyoto Protocol. The Kyoto Protocol established a targets-and-timetables approach to reducing global greenhouse gas emissions. The defining feature of the Kyoto Protocol was the framework it created that obligated developed country parties (generally referred to as the Annex I Parties) to collectively reduce their greenhouse gas emissions to 5 percent below 1990 levels by the time of the first compliance period, 2008–2012. As will be discussed in greater detail in Part C, based upon the collective 5 percent reduction goal, the Kyoto Protocol established independent legally-binding emissions reductions obligations for developed country parties, created a series of market-based mitigation tools to facilitate

---

[9] UNFCCC, *Status of Ratification of the Convention*, https://unfccc.int/essential_background/convention/status_of_ratification/items/2631.php.

[10] UNFCCC, *supra* note 1, Preamble.

[11] UNFCCC, *supra* note 1, art 4(1)(B).

implementation of these goals, and generally added further contour to the legal framework established by the UNFCCC.

As discussed in more detail later, the Protocol drew a sharp distinction between Annex I countries (virtually all developed countries) and Annex II countries (all of which were developing countries). Only the Annex I countries are subject to quantitative emissions targets under the Protocol.

Although the Kyoto Protocol was adopted in 1997, it did not come into force and thus did not take on a legally binding nature until 2005. This long lag was due to a number of factors. Primary among these was that the Kyoto Protocol contained a provision that required significant participation on the part of the developed country parties as a precondition to entry into force. Article 25 specified that the Protocol would not enter into force until: "not less than 55 Parties to the Convention, incorporating Parties included in Annex I which accounted in total for at least 55 per cent of the total carbon dioxide emissions for 1990 of the Parties included in Annex I."[12] This effectively meant that no matter how many developing countries ratified the treaty, the treaty would not enter into force unless and until the Parties to the Treaty included the majority of Annex I, i.e., developed counties. More specifically, because the United States accounted for roughly a quarter of global greenhouse gas emissions in 1990, either the United States plus a small handful of other developed countries had to ratify the treaty, or if the United Stated chose not to ratify the treaty, virtually *all* other major developed countries (e.g., the EU, Japan, Russia, Australia, Canada) would have to ratify the treaty in order to bring it into force. This provision, coupled with the fact that the treaty excluded all non-Annex I countries from legally binding emissions reduction obligations created great tension and highlighted the fact that global climate change negotiations were both highly political and materially shaped by a small but powerful handful of state actors.

C. Key State Actors in the Development of International Climate Change Law

From an early point in international climate change negotiations, a limited number of states have played a pivotal role in shaping the regime. During the early years of international climate change negotiations, the United States was one of the most influential state actors. In 1997, for example, through the leadership of Vice President Al Gore, the United States served as one of the key architects of the terms of the Kyoto Protocol. At the time, the United States was also the largest net global greenhouse gas emitter.

---

[12] Kyoto Protocol, *supra* note 2, art 25(1).

Accordingly, climate negotiators knew that the United States' response to the Protocol was critical. If the United States ratified the Protocol, then three primary results would follow: (1) it would be easier to bring the Protocol into force; (2) the largest global polluter would agree to meaningful emissions reductions, thus significantly limiting free rider problems and offering the real possibility of net global emissions reductions; (3) the international climate change regime would receive an important boost of support from a crucial actor. The United States' participation, thus, could help legitimize the institution and create added incentives for both developed and developing nations to engage in the process moving forward.

Yet, as was evident by the time of the 1997 treaty negotiations, U.S. ratification of the Protocol seemed unlikely. Although the Vice President aggressively supported the Kyoto Protocol, climate change was becoming an increasingly divisive issue back home. Prior even to the adoption of the Protocol, the Senate expressed opposition to any agreement that would harm the domestic economy or put the United States at an economic disadvantage in relation to its key economic competitors, including China and many of the other rapidly developing economies that had been excluded from legally binding emissions reductions obligations under the terms of the Protocol.[13]

The United States was not the only key actor, however. During the negotiations for the Protocol, the European Union emerged as a powerful proponent for aggressive action on climate change, including ratification and implementation of the Protocol. Once it became clear that the United States did not intend to ratify the Protocol,[14] the EU and other parties to the Protocol worked to convince Japan, Russia, and Australia to ratify the Protocol. Ultimately, these efforts were successful, and the Kyoto Protocol finally entered into force in 2005.

During the interim period between adoption and ratification of the Protocol, however, the terrain of international climate change politics underwent substantive changes. These change are critical to understanding the complex setting that defines contemporary climate change negotiations.

---

[13] Byrd-Hagel Resolution, S. Res. 98, 105th Cong., Rep. No. 105-54 (1997) (enacted). On July 25, 1997, the U.S. Senate passed the Byrd-Hagel Resolution by a margin of 95–0. This Resolution expressed the view of the Senate that the United States should not be a signatory to *any* protocol that exempted developing countries from legally binding obligations. The passage of the Byrd-Hagel Resolution virtually precluded the possibility that the United States subsequently would ratify the drafted Kyoto Protocol. *Id.*

[14] President Clinton never sent the Kyoto Protocol to the Senate floor. Upon his accession to the Presidency in 2001, President George W. Bush withdrew U.S. support for the Protocol, thus ruling out any possibility of U.S. ratification of the agreement in the near term.

Between 1997 and 2005, the greenhouse gas emissions footprint of the rapidly developing economies grew dramatically. Emissions in China, in particular, increased exponentially beginning in 2002. China's emissions grew so quickly that, by 2006, it had surpassed the United States as the largest annual net emitter of greenhouse gas emissions. This trend continued and, for the period between 1990–2011, not only was China the second largest cumulative emitter after the United States[15], but also, of the top ten emitters, five were developing countries.[16]

By as early as 2007, the role of the rapidly developing economies had become one of utmost importance. In 2007, the U.S. Energy Information Administration (EIA) projected that "by 2030, carbon dioxide emissions from China and India combined are projected to account for 31 percent of total world emissions."[17] In the wake of unprecedented rates of emissions growth in the developing economies, the global community had to adjust to a new reality: achieving meaningful emissions reductions could not be accomplished without full engagement on the part of the rapidly developing economies. As a result of these changes, the already tense debate over the relative roles and responsibilities of developed and developing countries in addressing climate change became even more of a pivotal question in ongoing negotiations.[18]

D. The Evolution of the Climate Change Regime: Copenhagen and Beyond

As these changes played out, international climate change negotiations continued from year to year. The Parties to the UNFCCC and the Kyoto Protocol made meaningful progress in developing discrete elements of the evolving climate change regime, including refining existing mitigation, adaptation, and financing tools. Yet, tensions deepened over the best way to achieve meaningful global greenhouse gas emission reductions and the overarching legal regime remained fairly static. As the end of the first Kyoto Compliance period approached, the future of the international

---

[15] For the period, 1990–2011, the United States accounted for 16 percent, and China for 15 percent of global emissions. *See* World Resources Institute, *6 Graphs Explain the World's Top Emitters*, http://www.wri.org/blog/2014/11/6-graphs-explain-world's-top-10-emitters (Nov. 24, 2014).

[16] The five developing countries ranking in the top ten include, in order: China, Brazil, Indonesia, India, and Mexico. *See* World Resources Institute, *supra*.

[17] Energy Info. Admin., U.S. Dep't of Energy, DOE/EIA-0484(2007), International Energy Outlook 2007 74–86 (2007).

[18] Alongside key state actors such as the United States, the European Union, and China, as the field of climate change law has matured, the pool of influential actors has increased to include a diverse array of state and regional partnerships, international organizations, and civil society groups spanning the public and private sectors. This phenomenon will be discussed further in Chapters 7 and 8.

climate change regime was uncertain. This period of uncertainty culminated in 2009 at the Copenhagen Climate Change Conference, the 15th meeting of the UNFCCC COP and the 5th meeting of the meeting of the parties (MoP) to the Kyoto Protocol.

Following multiple years of stalled negotiations, 2009 heralded signs of progress. Barack Obama assumed the Presidency of the United States on a platform that promised to prioritize domestic efforts to address climate change, and the new administration followed up with small steps to modify U.S. climate change law and policy. As the 2009 meeting neared, the European Union, the United States, China, India, and Brazil all offered evidence that they were willing to come to the negotiating table. The Copenhagen meeting represented the first time since the meeting in Kyoto in 1997, that all major emitters—which now included a much more diverse set of actors—would attend and actively engage in climate change negotiations.

The Copenhagen meeting, however, proved to be a tumultuous event wherein efforts to negotiate a new climate change agreement hit one roadblock after another. Power struggles between key states coupled with the existence of a multitude of varied voices and a mandate that required that decisions be made by consensus undermined efforts to agree upon a common legal or political architecture. One distinct aspect of the Copenhagen meeting ultimately allowed progress to be made, however. Unlike preceding COPs where the negotiations were largely left to lower-level diplomats and ministers, in Copenhagen, this model changed as a result of the presence and participation of Heads of State from the United States, Europe, China, India, South Africa, Brazil and elsewhere.

While the Copenhagen meeting did not produce a new multilateral legal instrument, what it did do was charter the course that now defines international climate change negotiations. With no collective agreement emerging, Heads of State from the United States, China, India, Brazil, and South Africa met in isolation and drafted an agreement. This agreement, the Copenhagen Accord,[19] is a three-page document that commits parties to continuing efforts to facilitate long-term cooperative action to combat climate change and stabilize greenhouse gas emissions with the goal of holding the increase in global temperature below 2° Celsius. It also called for the development of new and additional forms of funding, the creation of a new Green Climate Fund and a new Technology Mechanism, as

---

[19] U.N. Framework Convention on Climate Change Conference of the Parties, Copenhagen, Denmark, Dec. 7–19, 2009, *Report of the Conference of the Parties on its Fifteenth Session*, FCCC/CP/2009/11/Add.1 4 (March 30, 2010), http://unfccc.int/resource/docs/2009/cop15/eng/11a01.pdf.

well as for enhanced action on adaptation.[20] In its final provision, the Agreement calls for an "assessment of the implementation of this Accord to be completed by 2015."[21]

When the Copenhagen Accord was introduced to the COP in Copenhagen, it received mixed responses. A handful of developing countries, including Venezuela, Bolivia, Ecuador, opposed adopting the Accord as a formal UNFCCC COP decision. In the end, the COP agreed to "take note" of the Copenhagen Accord as a way to acknowledge its existence. The Copenhagen Accord, technically, is a marginal document of virtually no legal consequence. Yet, the document has played an important role in shaping the negotiations for a post-Kyoto legal agreement.

Most importantly, the Copenhagen Accord initiated a process by which all countries, whether developed or developing, would voluntarily pledge mitigation actions. This process represented the first time that developing countries were poised to make emission reduction commitments under the umbrella of the UNFCCC and it became the model for mitigation negotiations in the years that followed. Beyond mitigation, the Accord's emphasis on adaptation signaled a shift in post-2009 negotiations, with efforts to develop adaptation and mitigation strategies now receiving similar levels of international attention. Similarly, as mentioned earlier, the Copenhagen Accord called for the creation of a new "Green Climate Fund" to be established in order to serve as the operating entity of the financial mechanism for the Convention. At later meetings, the COP followed through in creating the Green Climate Fund.

At the subsequent COP meeting in Cancun (COP-16/MOP-6), the delegates picked up on the themes of the Copenhagen Accord and negotiated a set of agreements, *The Cancun Agreements,* that helped to resuscitate the viability of the global climate change institutions and begin the process of institutional reassessment.[22] In Cancun, the essence of the Copenhagen Accord was formalized and expanded on with respect to monitoring mitigation efforts, developing the principles of the new Green Climate Fund, creating a new mechanism to facilitate enhanced technology development and

---

[20] Rob Fowler, *Analysis of the Copenhagen Accord: An Initial Assessment of the Copenhagen Outcomes*, Teaching Climate/Energy Law & Policy, http://www.teachingclimatelaw.org/analysis-of-the-copenhagen-accord/.

[21] *See* FCCC/CP/2009/11/Add.1, *supra* note 19, at 7, ¶ 12.

[22] *See generally, Decision 1/CP.16, The Cancun Agreements: Outcome of the Work of the Ad Hoc Working Group on Long-Term Cooperative Action under the Convention,* (2010), FCCC/CP/2010/7/Add.1 http://unfccc.int/files/meetings/cop_16/application/pdf/cop16_lca.pdf; *The Cancun Agreements: Outcome of the work of the Ad Hoc Working Group on Further Commitments for Annex I Parties under the Kyoto Protocol at its Fifteenth Session,* (2010), FCCC/KP/CMP/2010/12/Add.1, http://unfccc.int/files/meetings/cop_16/application/pdf/cop16_kp.pdf.

transfer, and establishing the Cancun adaptation framework to facilitate action on adaption, including the creation of a new Adaptation Committee. In key part, the Cancun Agreements reaffirmed the need to achieve cuts in greenhouse gas emissions "deep enough to hold the increase in global average temperature below 2°C above preindustrial levels."[23]

The next year, in 2011, in Durban, South Africa, the COP to the UNFCCC adopted the Durban Platform for Enhanced Action (the Durban Platform).[24] The Durban Platform launched a new round of negotiations aimed at developing "a protocol, another legal instrument or an agreed outcome with legal force" for the period from 2020 onward. The Durban Platform directs the parties to the Convention to conclude negotiations for the new instrument by 2015.

One critical point of departure between the Durban round of negotiations and original negotiations for the UNFCCC is the emphasis on the legal nature of the resulting instrument. The 1990 U.N. General Assembly Resolution—which ultimately resulted in the negotiations of the UNFCCC—called for a framework convention (i.e., a legal instrument). In contrast, the Durban Platform called for a more generalized process to develop a "protocol, another legal instrument or a legal outcome," and omitted any language about legally binding commitments. This shift was indicative of a larger trend in international climate change law, which reflects growing recognition that a new, more flexible, inclusive, and facilitative approach is needed to mobilize global cooperation on climate change.

Between the Durban COP in 2011 and the Paris meeting in 2015, the parties to the UNFCCC and the Kyoto Protocol made progress on developing new funding structures, maintaining existing mitigation strategies,[25] fulfilling the new mandate that specified that "[a]daptation must be addressed with the same priority as mitigation and requires appropriate institutional arrangements to enhance adaptation action and support",[26] and otherwise making progress

---

23 *Id.* at para. 4.

24 UNFCCC, FCCC/CP/2011/L.10, *Establishment of an Ad Hoc Working Group on the Durban Platform for Enhanced Action*, Draft Decision -/CP.17, (Dec. 10, 2011).

25 UNFCCC, *Outcome of the Work of the Ad Hoc Working Group on Further Commitments for Annex I Parties under the Kyoto Protocol at its Sixteenth Session*, Draft Decision -/CMP.7 (2011), http://unfccc.int/files/meetings/durban_nov_2011/decisions/application/pdf/awgkp_outcome.pdf. Similarly, the Parties to the Kyoto Protocol reaffirmed the continuing validity of the Clean Development Mechanism (CDM), one of the most important creations of the Kyoto Protocol. *See* UNFCCC, Draft decision -/CMP.7, Further Guidance Relating to the Clean Development Mechanism, http://unfccc.int/files/meetings/durban_nov_2011/decisions/application/pdf/cmp7_cdm guidance.pdf.

26 *Decision 1/CP.16, The Cancun Agreements: Outcome of the Work of the Ad Hoc Working Group on Long-Term Cooperative Action under the Convention* at Part I(2)(b). *See also* UNFCCC, *Outcome of the Work of the Ad Hoc Working Group on Long-*

with respect to developing and operationalizing key aspects of the institutional structure that would be needed to help implement any new international climate agreement.[27]

With respect to the nature of the future agreement, at the COP in Warsaw in 2013, through the decision on *Further Advancing the Durban Platform*, the parties to the Convention first created the parameters for what would become the Paris Agreement, including the Intended Nationally Determined Contributions (INDCs) model. One year later, at the COP in Lima in 2014, through the *Lima Call for Climate Action*, a new bottom-up, inclusive mitigation model was formalized as the parties reiterated the "invitation to each party to communicate to the secretariat its intended nationally determined contribution towards achieving the objective of the Convention",[28] and specified that each party's intended nationally determined contribution towards achieving the objective of the Convention, as set out in Article 2, should represent a progression beyond the current undertaking of that party and "shall address in a balanced manner, inter alia, mitigation, adaptation, finance, technology development and transfer, and capacity-building, and transparency of action and support."[29]

Leading up to the 2015 meeting in Paris, the parties to the UNFCCC proposed shifting from a model of top-down, overarching mitigation commitments to a model premised on individual states determining how they would contribute to efforts to limit climate change. In the run-up to the Paris meeting, all major emitters, including the United States, China, India, and the European Union heeded the call in the *Lima Call for Climate Action* and released INDCs in which they set out the parameters for what they would be willing to do to address climate change. The willingness of all major emitters to engage in this process created positive momentum for international agreement leading into the 2015 meeting in Paris.

Term Cooperative Action under the Convention, Draft decision [-/CP.17], Part III, http://unfccc.int/files/meetings/durban_nov_2011/decisions/application/pdf/cop17_lca outcome.pdf#page=17; UNFCCC, *National Adaptation Plans*, Draft decision -/CP.17, FCCC/CP/2011/L.8/Add.1 (Dec. 10, 2011).

[27] *See, e.g.*, UNFCCC, *Revision of the UNFCCC reporting guidelines on annual inventories for Parties included in Annex I to the Convention*, Draft decision -/CP.17 (2011), http://unfccc.int/files/meetings/durban_nov_2011/decisions/application/pdf/cop 17_annual_inventories.pdf.

[28] Lima Call for Climate Action, ¶ 10, Dec. 1/CP.20 (Dec. 14, 2014), *in* COP Report No. 20, Addendum, at 2, UN Doc. FCCC/CP/2014/10/Add.1 (Feb. 2, 2015).

[29] *Id.*

E. The 2015 Paris Agreement: A New International Framework

In December of 2015,[30] the Parties to the UNFCCC successfully negotiated a new successor agreement to the Kyoto Protocol. This new agreement, the Paris Agreement, represents the culmination of efforts, begun in Copenhagen in 2009, to create a more flexible and bottom-up model for addressing climate change. Daniel Bodansky describes the result as "a Goldilocks solution that is neither too strong (and hence unacceptable to key states) nor too weak (and hence ineffective)."[31]

The Paris Agreement commits the Parties to "holding the increase in the global average temperature to well below 2°C above pre-industrial levels and pursuing efforts to limit the temperature increase to 1.5°C above pre-industrial levels," to increasing pathways towards adaptation, and to improving climate finance—all within a framework focused on reflecting "equity and the principle of common but differentiated responsibilities and respective capabilities, in the light of different national circumstances."[32] The primary way the Agreement envisions meeting these goals is through the submission of Nationally Determined Contributions (NDC) that reflect each Parties' highest possible ambition within the common but differentiated responsibilities (CBDR) framework, with each round of NDC submissions envisioned as being progressively ambitious.

The underlying role of NDCs is for Parties to demonstrate a commitment to helping meet the 2°C goal and for the global community, collectively, to understand how all of the NDCs combine and intersect to allow progress towards that goal. Through the NDC process, Parties define their ambition for emissions reduction. Pursuant to the *Lima Call for Climate Action*, mitigation goals in the NDCs "may include, as appropriate, . . . quantifiable information on the reference point (including, as appropriate, a base year), time frames and/or periods for implementation, scope and coverage, planning processes, assumptions and methodological approaches including those for estimating and accounting for anthropogenic greenhouse gas emissions."[33] Beyond setting out mitigation commitments, Parties may also use the NDC process to define how they plan to "adapt to climate change impacts, and what support they need from, or will provide to, other countries to adopt low-carbon

---

[30] The Paris Agreement entered into force on November 4, 2016. As of June 2021, 193 Parties, of the 197 Parties to the UNFCCC, have ratified the Paris Agreement.

[31] Daniel Bodansky, *The Paris Climate Change Agreement: A New Hope?*, 110 AM. J. INT'L L. 288, 289 (2016).

[32] Paris Agreement, *supra* note 3, at Art 2.

[33] Lima Call for Climate Action, *supra* note 28, ¶ 14.

pathways and to build climate resilience."[34] The NDCs are flexible tools that allow nations to communicate their overarching resource needs and climate goals.[35]

Pursuant to the Lima Call for Climate Action, the two key components that every NDC should aim to embody are ambition and equity.[36] Specifically, Parties are tasked with describing why they consider their nationally determined contributions to be fair and ambitious.[37] In this way, fairness and ambition become the twin pillars for assessing the adequacy of Party submissions, with ambition relating to mitigation effectiveness and fairness, presumably, relating to evolving notions of equity.

The level of ambition of NDCs can be specifically keyed to the objective, as set out in the Paris Agreement, to limit the increase in the global average temperature to well below 2°C. Collective ambition, therefore, is "lacking when the aggregate policies and actions of all countries are deemed insufficient to meet the 2°C goal."[38] Individual Party NDCs, on the other hand, should be viewed within a context of common but differentiated responsibilities and respective capacities—that is, is the country's commitment to reducing (or peaking, or otherwise limiting) greenhouse gas emissions in line with their respective position within the global community? This assessment, of course, is complicated by the

---

[34] *What is an INDC?*, World Resources Institute, http://www.wri.org/indc-definition.

[35] *See id.*

[36] A third important element is transparency. Transparency is important "so that stakeholders can track progress and ensure countries meet their stated goals" and can be achieved, in part, through making the NDCs public, thus providing the global community and other interested parties with the ability to track progress and ensure countries meet their stated goals. Of course, publishing an NDC is only one element of transparency; transparency also requires continuing access to information in order to track compliance, or implementation of proposed strategies. *What is an INDC?*, *supra* note 34.

[37] Lima Call for Climate Action, *supra* note 28, ¶ 14. I n full, paragraph 14 states:

> Agrees that the information to be provided by Parties communicating their intended nationally determined contributions, in order to facilitate clarity, transparency and understanding, may include, as appropriate, inter alia, quantifiable information on the reference point (including, as appropriate, a base year), time frames and/or periods for implementation, scope and coverage, planning processes, assumptions and methodological approaches including those for estimating and accounting for anthropogenic greenhouse gas emissions and, as appropriate, removals, and how the Party considers that its intended nationally determined contribution is fair and ambitious, in light of its national circumstances, and how it contributes towards achieving the objective of the Convention as set out in its Article 2.

Id.

[38] Edward Cameron, *What is Ambition in the Context of Climate Change?*, World Resources Institute, (November 26, 2012), https://www.environmental-expert.com/articles/what-is-ambition-in-the-context-of-climate-change-331467.

indeterminacy surrounding understandings of CBDR, as discussed *infra* in Part II(A), and the interplay between CBDR and equity, but suggests, at a minimum, that a country's level of ambition should be assessed based on its particular historic, economic, political, and environmental attributes. Other factors relevant to the assessment of ambition include the real and anticipated level of implementation of climate policies, as well as "the finance, capacity building, and technology transfer support offered to help developing nations—arguably the countries that are most vulnerable to climate change—mitigate and adapt to global warming's impacts."[39]

There is significant variety among the NDCs in terms of length, detail, and commitment. Given that the invitation to submit NDCs in the Lima Call for Climate Action contained broad language about the content of the INDCs and that the Paris Agreement did little to narrow or add detail to the NDC format, it should not come as a surprise that countries have adopted very different approaches to structuring their NDCs, approaches that are both strategic and, often, reflective of the approaches different states have taken during the climate negotiations.

As brief examples, the first United States NDC is five pages long, contains concise and precise overviews of what the country is willing to do and why; there is no excessive detail or justification offered for what and why the United States commits to doing. At the heart of the United States' NDC is an intention "to achieve an economy-wide target of reducing its greenhouse gas emissions by 26–28 per cent below its 2005 level in 2025 and to make best efforts to reduce its emissions by 28%."[40] Like that of the United States, the first submission of the European Union is concise, running less than five pages, and precise about its global commitment. Dissimilar to the United States, the European Union offers a bit more insight into how it believes its NDC to be fair and, thus, what fairness might mean. To begin, the European Union offers an economy-wide absolute emissions reduction commitment based on a 1990 baseline. Specifically, the EU and its Member States commit "to a binding target of an at least 40% domestic reduction in greenhouse gas emissions by 2030 compared to 1990, to be fulfilled jointly."[41]

China, the largest global emitter of greenhouse gas emissions, framed its first NDC with the understanding that "China is among those countries that are most severely affected by the adverse

---

[39] *Id.*

[40] USA NDC, https://unfccc.int/sites/default/files/NDC/2022-06/U.S.A.%20First%20NDC%20Submission.pdf.

[41] Submission by Latvia and the European Commission on Behalf of the European Union and its Member States 1 (March 6, 2015), https://unfccc.int/sites/default/files/LV-03-06-EU%20INDC.pdf.

impacts of climate change" and, thus, is driven to address climate change not only by its own domestic economic and sustainability needs but also "by its sense of responsibility to fully engage in global governance, to forge a community of shared destiny for humankind and to promote common development for all human beings."[42] The NDC then sets out both mitigation and adaptation goals, including peaking CO_2 emissions around 2030 and improving disaster prevention and reduction.[43]

In contrast to some of the more concise submissions, India's first NDC was 38 pages long and offered a highly detailed, intricately contextualized picture of India's economic and environmental history and uses this detailed picture to frame its NDC, which does not begin until page 29 of the document.[44] India's NDC differs from that of the United States and the European Union in that it does not focus on absolute emissions reductions, and it differs from China's NDC in that it does not seek to peak emissions by a particular point in time. India's NDC instead focuses on limiting the growth of fossil fuels primarily through the shift towards reduced emissions intensity and greater reliance on clean energy and climate friendly development, but it also differs from the United States, the European Union and, even, China in its emphasis on internal adaptation, capacity building, and healthy and sustainable living as core components of its international commitment.

With respect to some of the most vulnerable states, there are a number of submissions from small island states such as Tuvalu, Nauru, and Niue that run roughly 9–10 pages.[45] These submissions offer relatively straightforward and comparatively ambitious mitigation goals, but contextualize those goals carefully within the

[42] China, *Enhanced Actions on Climate Change: China's Intended Nationally Determined Contributions*, https://unfccc.int/sites/default/files/NDC/2022-06/China%27s%20First%20NDC%20Submission.pdf.

[43] China's full list of mitigation and adaptation goals is as follows:

> To achieve the peaking of carbon dioxide emissions around 2030 and making best efforts to peak early; To lower carbon dioxide emissions per unit of GDP by 60% to 65% from the 2005 level; To increase the share of non-fossil fuels in primary energy consumption to around 20%; and To increase the forest stock volume by around 4.5 billion cubic meters on the 2005 level. Moreover, China will continue to proactively adapt to climate change by enhancing mechanisms and capacities to effectively defend against climate change risks in key areas such as agriculture, forestry and water resources, as well as in cities, coastal and ecologically vulnerable areas and to progressively strengthen early warning and emergency response systems and disaster prevention and reduction mechanisms.

Id. at 5.

[44] India's Nationally Determined Contribution, https://unfccc.int/sites/default/files/NDC/2022-06/INDIA%20INDC%20TO%20UNFCCC.pdf.

[45] The submissions of some other small island states are longer, such as Kiribati's NDC, which is 27 pages long. NDC Kiribati, https://unfccc.int/sites/default/files/NDC/2022-06/INDC_KIRIBATI.pdf.

context of the insignificant role that they play in contributing global greenhouse gas emissions and the existential threat they face if global climate change is not limited and adaptation and loss and damage is not prioritized.

These examples represent just a handful of the approaches that Parties take to crafting their NDCs. What is evident is that the NDCs represent much more than simple statement of what Parties are willing to do to address climate change, they represent an opportunity for Parties to share with the global community, either directly or indirectly, how climate change shapes the past, present, and future of their economic, political, and sovereign well-being and what this means for individual and collective mitigation and adaptation goals.

Beyond the NDCs, the Paris Agreement also affirms the importance of adaptation and places greater emphasis on adaptation planning, loss and damage mechanisms, climate finance, inclusive mitigation mechanisms, and other measures linked to efforts to promoting equity and fairness in climate actions. In particular, Article 5 encourages countries to participate in efforts to reduce deforestation, the most important source of greenhouse gases after energy production. We discuss these efforts in Chapter 8. In addition, the Agreement mandated a global stocktake, or a process whereby beginning in 2023, and every five years thereafter, countries must assess their collective progress towards reaching the goals of the Paris Agreement. Subsequently, each country's NDC must be "informed by the outcomes of the global stocktake" to ensure continuing progress towards achieving the long-term goals of the Agreement.[46]

Although the Paris Agreement represented a positive step forward in international cooperation, even if Parties fully fulfilled the commitments they make in their first NDCs, it is unlikely that this would hold warming below 2°C, much less achieving the 1.5°C target; in fact the Parties noted with concern "that the estimated aggregate greenhouse gas emission levels in 2025 and 2030 resulting from the intended nationally determined contributions do not fall within least-cost 2°C scenarios but rather lead to a projected level of 55 gigatonnes in 2030."[47] Of course, the hope was that these initial NDC

---

46 Paris Agreement, *supra* note 3, at Art 4(9).

47 Report of the Conference of the Parties on its twenty-first session, held in Paris from 30 November to 13 December 2015 II(17), UNFCCC, FCCC/CP/2015/10/ Add.1 (Jan. 29, 2016) (In key part the decision "[n]otes *with concern* that the estimated aggregate greenhouse gas emission levels in 2025 and 2030 resulting from the intended nationally determined contributions do not fall within least-cost 2°C scenarios but rather lead to a projected level of 55 gigatonnes in 2030, and *also notes* that much greater emission reduction efforts will be required than those associated with the intended nationally determined contributions in order to hold the increase in

targets would be followed by stricter ones that would bring the world closer to a safe outcome.[48]

Following the adoption of the Paris Agreement, President Trump called into question its potential effectiveness when, on June 1, 2017, he announced that the United States would be withdrawing from the agreement. More precisely, he declared that "as of today, the United States will cease all implementation of the non-binding Paris Accord and the draconian financial and economic burdens the agreement imposes on our country."[49] This decision did not have any immediate legal effect. Article 28 of the Paris Agreement provides that "[a]t any time after three years from the date on which this Agreement has entered into force for a Party, that Party may withdraw from this Agreement by giving written notification." The withdrawal then takes effect "expiry of one year from the date of receipt . . . or on such later date as may be specified in the notification of withdrawal." Thus, the U.S. was not able to give formal notification of its withdrawal until November 4, 2020, three years after the agreement went into force.

The symbolic and substantive effects of President Trump's announcement, however, were significant. First, it set the tone for the Trump Administration's approach to climate change. That tone being total defiance of the idea that the United States should be a cooperative actor on climate change. Second, the decision to cease implementation of the Agreement had immediate effect on global mitigation and adaptation efforts, given that it meant that the United States—the second largest-global GHG emitter—would no longer commit to reducing its emissions in line with the pledge it made under the Agreement through its first NDC. Third, it also meant that the United States would immediately stop providing the climate finance that it had committed to under the Obama Administration, with the effect of weakening global mitigation and adaptation efforts.

---

the global average temperature to below 2 °C above pre-industrial levels by reducing emissions to 40 gigatonnes or to 1.5 °C above pre-industrial levels by reducing to a level to be identified in the special report referred to in paragraph 21 below.)

[48] As one positive indication of the impact of evolving international negotiations and the translation of this into state action, over the past few years, several European states have adopted climate laws that courts have subsequently found the state to be in violation of—for lack of ambition—on grounds of constitutional and human rights violations. These state-based actions are indicative of downward pressure on the development of climate law. *See, e.g.*, Case C-565/19 P, *Armando Ferrão Carvalho and Others v. The European Parliament and the Council*, ECLI:EU:C:2021:252 (Mar. 25, 2021); *Notre Affaire à Tous and Others v. France*, N°s 1904967, 1904968, 1904972, 1904976/4-1 (Oct. 14, 2021); *Urgenda v. the Netherlands* (Supreme Court, 20 December 2019) ECLI:NL:HR:2019:2007.

[49] President Donald J. Trump, Statement by President Trump on the Paris Climate Accord (June 1, 2017), https://perma.cc/U9YF-9792.

President Trump's announcement was met with considerable defiance at both the domestic and international levels. At the domestic level, his announcement was met with an immediate outpouring of resistance and widespread efforts to mobilize subnational and non-state actors to step into the void to help keep the United States on track to pursuing domestic and international commitments to address climate change. On the same day that President Trump announced the United States' de facto withdrawal from the Paris Agreement, the governors of California, Washington, and New York announced they had formed a new partnership, the United States Climate Alliance, aimed at advancing the goals of the Paris Agreement and fulfilling the United States' obligations thereunder.[50] In addition, eighty-four U.S. mayors, representing forty million Americans, issued a joint statement declaring their intention to "adopt, honor, and uphold the commitments to the goals enshrined in the Paris Agreement."[51] Complementing the state and city initiatives, in June 2017, a group of mayors, governors, and business leaders launched the "We Are Still In" initiative that brought together a bipartisan coalition of "mayors, county executives, governors, tribal leaders, college and university leaders, businesses, faith groups, cultural institutions, healthcare organizations, and investors," declaring their intent to continue efforts to implement the United States international climate pledge.[52] Also in June 2017, California Governor, Jerry Brown, together with Michael Bloomberg launched "America's Pledge," an initiative to "compile and quantify the actions of states, cities and businesses in the United States to drive down their greenhouse gas emissions consistent with the goals of the Paris Agreement."[53]

At the international level, President Trump's announcement was met with reactions varying from a symbolic shrug to exasperated defiance. France joined with Germany and Italy in a statement taking note "with regret of the decision by the United States of

---

[50] U.S. CLIMATE ALLIANCE, https://www.usclimatealliance.org [https://perma.cc/G7F3-HR23]. By summer 2019, 25 governors were members. Press Release, U.S. Climate Alliance, Montana Governor Steve Bullock Becomes 25th Governor to Join U.S. Climate Alliance (July 1, 2019), https://www.usclimatealliance.org/publications/2019/7/1/montana-governor-steve-bullock-becomes-25th-governor-to-join-us-climate-alliance [https://perma.cc/A6SP-J5KA].

[51] *Paris Climate Agreement: 407 US Climate Mayors Commit to Adopt, Honor and Uphold Paris Climate Agreement Goals*, CLIMATE MAYORS, https://climatemayors.org/actions-paris-climate-agreement/ [https://perma.cc/2WED-J7QB]. As of July 2019, the pact included 407 U.S. Mayors representing 70 million Americans. *Id.*

[52] *"We Are Still In" Declaration*, WE ARE STILL IN, https://www.wearestillin.com/we-are-still-declaration [https://perma.cc/5B55-AZPU].

[53] Press Release, Office of Governor Edmund G. Brown, California Governor Jerry Brown and Michael Bloomberg Launch "America's Pledge" (July 12, 2017), https://ca.gov/archive/gov39/2017/07/12/news19872/index.html [https://perma.cc/K3U8-DYEE].

FRANCE, GERM, + ITALY

America to withdraw from the universal agreement on climate change," and committing to "step up efforts to support developing countries, in particular the poorest and most vulnerable, in achieving their mitigation and adaptation goals."[54] The Prime Ministers of Canada and India expressed similar frustration and resolve.[55] Perhaps, most importantly, preceding and following President Trump's announcement, the Chinese government continued to express support for the Paris Agreement and disappointment in the United States' efforts to undermine the global pact. Thus, while President Trump's effective withdrawal from the Paris Agreement undercut international climate efforts, high levels of domestic and international support for the Paris Agreement persisted and on January 20, 2021, on his first day in office, President Biden officially rejoined the Paris Agreement.[56]

F. The 2021 Glasgow Climate Pact

Following the adoption of the Paris Agreement, the Parties began the process of developing a set of implementation rules for the Agreement. These efforts culminated at COP24 in Katowice, Poland in 2018, with the release of a set of implementation rules commonly referred to as the Paris rulebook. The rulebook established key guidelines for the development of NDCs; adaptation communications; loss and damage reporting and information gathering; climate finance; transparency; and the global stocktake. Notably, at the time, the Parties to the Agreement were unable to agree on rules for the development of the Article 6 mechanisms, which were finally developed at COP26 in Glasgow in 2021.

Following a delay in negotiations due to the global coronavirus pandemic,[57] on November 13, 2021, the Parties to the Agreement signed the Glasgow Climate Pact. The Glasgow Climate Pact represents the most significant development in international climate change law since the Paris Agreement. Although it is a non-binding

---

[54] *Statement by Italy, France, and Germany on the US Withdrawal from the Paris Agreement on Climate*, AMBASCIATA D'ITALIA, WASH. D.C. (June 1, 2017), https://ambwashingtondc.esteri.it/ambasciata_washington/en/sala-stampa/dall_ambasciata/2017/06/dichiarazione-italia-germania-francia.html [https://perma.cc/LB3Q-TBSW].

[55] *Statement by the Prime Minister of Canada in Response to the United States' Decision to Withdraw from the Paris Agreement*, PRIME MINISTER CAN. (June 1, 2017), https://pm.gc.ca/eng/news/2017/06/01/statement-prime-minister-canada-response-united-states-decision-withdraw-paris [https://perma.cc/5ECX-YE83]; PMO India (@PMO India), TWITTER (June 3, 2017, 5:29 AM), https://twitter.com/PMOIndia/status/870980871720845312 [https://perma.cc/BP5Q-QZA8].

[56] U.S. Department of State, *Press Statement: The United States Officially Rejoins the Paris Agreement* (Feb. 19, 2021), https://www.state.gov/the-united-states-officially-rejoins-the-paris-agreement/.

[57] COP 25 was held in 2019 in Madrid, Spain, but COP26, which was due to take place in Glasgow, Scotland in 2020 was delayed due to the global coronavirus pandemic. As a result, COP26 did not occur until November 2021.

instrument, it sets the path for accelerating implementation of the Paris Agreement. The Glasgow Climate Pact calls upon Parties to the Paris Agreement "to accelerate the development, deployment and dissemination of technologies . . . to transition towards low-emission energy systems" and "accelerat[e] efforts towards the phase-down of unabated coal power and phase-out of inefficient fossil fuel subsidies."[58] In addition, the Pact calls on Parties to accelerate their plans for emissions reductions over the ensuing decade, including through the development of new NDCs every five years.

The Glasgow Climate Pact did not fundamentally alter the terrain of international climate change law, but it kept the Paris Agreement ambition of containing warming to 1.5° alive and created a more defined pathway forward for Parties to develop increasingly ambitious NDCs. In addition, the Pact pushed climate finance forward by setting a goal for developed countries to double the funding provided to developing countries for adaptation action by 2025.

Notably, in the lead up to the COP26 meeting in Glasgow, the United States submitted a revised NDC stating that: "the United States is setting an economy-wide target of reducing its net greenhouse gas emissions by 50–52 percent below 2005 levels by 2030", which would put the United States on target to achieve net-zero emissions by no later than 2050.[59] Moreover, in anticipation of the meeting, the United States and China issued a first of its kind declaration to work cooperatively "to reduce emissions aimed at keeping the Paris Agreement-aligned temperature limit within reach."[60]

The US-China pact was one of various soft law and cooperative agreements released around the same time as COP26. Other notable multilateral cooperative agreements focused on: reducing human-caused methane emissions by 30 percent between 2020 and 2030, phasing out coal power and stopping the construction of new coal plans, halting and reversing forest loss and land degradation by 2030, increasing shares of sales of new cars/vans being zero

---

[58] The Pact is the first UNFCCC climate instrument to explicitly reference coal and fossil fuels, but it should be noted that an earlier draft of the instrument called for the "phase out" of the use of unabated coal, but this was altered to read the "phase-down" of unabated coal after India, with the support of China and other countries, objected to the language of "phase out".

[59] The United States of America, Nationally Determined Contribution, *Reducing Emissions in the United States: A 2030 Emissions Target* (April 21, 2021), https://unfccc.int/sites/default/files/NDC/2022-06/United%20States%20NDC%20April%2021%202021%20Final.pdf.

[60] U.S. Department of State, Media Note: *U.S.-China Joint Glasgow Declaration on Enhancing Climate Action in the 2020s*, (Nov. 10, 2021), https://www.state.gov/u-s-china-joint-glasgow-declaration-on-enhancing-climate-action-in-the-2020s/.

emissions to 100% by 2035/2040, stopping funding for fossil fuel projects abroad, and incentivizing new financial commitments from the private sector to facilitate the transition to a net-zero economy.[61]

Despite the ambitious nature of the Glasgow Climate Pact and supporting cooperative agreements, a United Nations report issued in advance of COP26 suggested that existing commitments are not enough to achieve the long-term goals of the Paris Agreement. Specifically, the report assessed the revised climate pledges the Parties submitted prior to COP26 and determined that the combined effect of these pledges and other mitigation measure would put the world on track for a global temperature rise of 2.7°C by the end of the century. That is well above the goals of the Paris climate agreement and would lead to catastrophic changes in the Earth's climate. To keep global warming below 1.5°C this century, the aspirational goal of the Paris Agreement, the world needs to halve annual greenhouse gas emissions in the next eight years.

While the report suggests that reducing methane and deploying carbon markets could help reduce emissions, these changes are far from guaranteed. Thus, even with strengthening efforts to reduce emissions, it may be necessary to use large-scale CO_2 removal later in this century, which would require the development of new technologies, which we discuss in Chapter 9.

II. General Principles of International Climate Change Law

International climate change negotiations past and present are guided by a set of general principles. These principles originate from the larger field of international environmental law but are made specific to climate change through the terms of the UNFCCC. The general principles that form the normative backbone of the climate regime place equity considerations at the center of the regime. In key part, the UNFCCC does the following: characterizes the Earth's climate system as of common concern to humankind; articulates the importance of protecting the climate system for present and future generations; recognizes that the common responsibility to protect the

---

[61] *See* Carbon Brief, *COP26: Key Outcomes Agreed at the UN Climate Talks in Glasgow* (Nov. 15, 2021), https://www.carbonbrief.org/cop26-key-outcomes-agreed-at-the-un-climate-talks-in-glasgow/. *See also* Glasgow Financial Alliance for Net-Zero (GFANZ), *Amount of Finance Committed to Achieving 1.5°C Now at Scale Needed to Deliver the Transition* (Nov. 3, 2021), https://www.gfanzero.com/press/amount-of-finance-committed-to-achieving-1-5c-now-at-scale-needed-to-deliver-the-transition/ ("commitments, from over 450 firms across 45 countries, can deliver the estimated $100 trillion of finance needed for net zero over the next three decades").

[61] UNFCCC, *supra* note 1, art 3(1). The principle of common but differentiated responsibilities and respected capacities is variously referred to as CBDR and CBDRRC. Here, we use the more concise version, "CBDR".

climate system should be differentiated among parties on the basis of capacity; promotes a precautionary approach to addressing climate change; recognizes the right to sustainable development; and emphasizes the importance of promoting an open international economic system. These principles are laid out clearly in the treaty. In practice, efforts to realize these principles in the evolving regime face persistent challenges.

Three principles play a particularly important role in climate negotiations. These include the principles of: (1) common but differentiated responsibilities, (2) intergeneration equity, and (3) the precautionary principle. Each of these will be discussed briefly.

A. Common but Differentiated Responsibility (CBDR)

Very few topics have generated as much debate or as much discord as the meaning and practical implications of implementing the CBDR principle. The UNFCCC provides that Parties to the Treaty "should protect the climate system . . . on the basis of equity and in accordance with their common but differentiated responsibilities and respective capabilities."[62] This provision has created one of the most enduring and practically important ethics-based climate debates: what does common but differentiated responsibilities mean and how should it be realized?

The practice of differentiating responsibilities among parties has existed in international law for years,[63] but the specific term 'common but differentiated responsibilities' was first formally used in 1992 in the Rio Declaration on Environment and Development,[64] and in the UNFCCC. In the UNFCCC, the term suggests that the international community shares a common responsibility for protecting the global climate system, but that responsibility should be differentiated among the countries of the world based upon their

---

[62] UNFCCC, *supra* note 1, art 3(1). The principle of common but differentiated responsibilities and respected capacities is variously referred to as CBDR and CBDRRC. Here, we use the more concise version, "CBDR".

[63] *See* Christopher Stone, *Common but Differentiated Responsibilities in International Law* (2004) 98 AJIL 276, 278. The 1972 Stockholm Declaration laid the foundations for CBDR in environmental law:

> Without prejudice to such criteria as may be agreed upon by the international community, or to standards which will have to be determined nationally, it will be essential in all cases to consider the systems of values prevailing in each country and the extent of the applicability of standards which are valid for the most advanced countries but which may be inappropriate and of unwarranted social cost for the developing countries.

Report of the United Nations Conference on the Human Environment (Stockholm, Sweden, 5–16 June 1972) (16 June 1972) UN Doc A/CONF.48/14 (Stockholm Declaration).

[64] United Nations Conference on Environment and Development, Rio de Janeiro, Braz., June 3–14, 1992, *Rio Declaration on Environment and Development*, Principle 7, U.N. Doc A/CONF. 151/26 (Vol. 1) (Aug. 12, 1992).

differing levels of capacity, arguably taking into account past contributions to the problem (i.e., historic greenhouse gas emissions), as well as present capacity to respond.[65]

The CBDR principle underpins the entire structure of the UNFCCC, the Kyoto Protocol, and the Paris Agreement. Both treaties are premised upon the need for multilateral cooperation to address a global problem that requires collective action. Both treaties also distinguish responsibility for responding to climate change between developed and developing countries, place greater responsibility on the shoulders of the developed country states—that is, the states that, in 1992 and 1997, were responsible not only for the highest levels of past and present emissions, but also were in a better economic position to limit the causes and consequences of climate change. This is not to say that CBDR is based on an expectation of uncontrolled emissions growth in developing countries, but it places the burden on developed countries to provide the technology and financing for developing country efforts.

The principle is informed by concerns over fairness, distributive justice, and practicality; yet many developed countries, including the United States, opposed the inclusion of CBDR in the international climate change treaties from the outset. Opposition to CBDR took many forms, but the most fundamental argument was that implementing the principle would create additional legal obligations and, thus, economic burden for a limited group of actors. This differentiation, the United States and others feared, could detrimentally affect domestic economies and create imbalances in international economic relations. Despite resistance, CBDR emerged as the "overall principle guiding future development of the regime."[66]

Long-standing disagreements over the proper way to interpret and implement the principle of CBDR have become increasingly divisive over time. In particular, the static interpretation of CBDR that places the burden of achieving emissions reductions on the developed countries (as defined in 1992 in Annex 1 of the UNFCCC), while excluding all developing countries, including the rapidly developing economies (e.g., China, India, and Brazil) from any type of legally-binding emissions reduction commitment, has come under increased fire over time. Not only does this stark interpretation potentially place developed countries in a position of economic disadvantage to many of their competitors, but it also threatens to undermine the collective action necessary to address climate change.

---

[65] Lavanya Rajamani, *The Principle of Common but Differentiated Responsibility and the Balance of Commitments under the Climate Regime,* 9 REV EUR COMMUNITY & INTL ENVTL L 120, 121 (2000).

[66] Rajamani, *supra* at 64.

Changing views of the meaning of CBDR create divisions that cut across developed-developing country lines. Developing countries differ far more in economic terms than they did in 1990 when the UNFCCC was negotiated. Developed countries, such as the United States and the European Union, alongside developing countries—including, in key part, the Alliance of Small Island States—converge around the view that any effective international climate regime necessarily must include emissions reductions commitments on the part of *all* of the world's biggest emitters. On the other hand, the rapidly developing economies, including China and India, push back against an expanded interpretation of CBDR.

The Paris Agreement reemphasizes the importance and centrality of the principle of CBDR and does not dramatically shift or seek to clarify existing understandings of CBDR. However, the move in the Lima Call for Climate Action to invite Parties to focus on fairness as a cornerstone of their INDCs disrupts the twenty-plus-year frame for state-based climate negotiations that centered on broad notions of equity and CBDR centered within a split-world frame.[67] The concept of fairness, however, remains poorly defined, and variously understood by different states. As a result, the extent and impact of this disruption will be played out in the coming years leading to continued debate about the how to understand CBDR in a contemporary context. The question of how the collective burden should be distributed remains one of the most contentious and unanswered questions in global climate change negotiations.

B. Intergenerational Equity

The principle of CBDR emphasizes the importance of accounting for differences among parties in the context of thinking about present and future responsibilities to respond to climate change. The UNFCCC prefaces the principle of CBDR by directing parties to act based upon a related set of equity considerations: concern for both present and future generations.[68] Specifically, Article 3 of the UNFCCC calls for Parties to the convention "to protect the climate system for the benefit of present and future generations of humankind, on the basis of equity".[69]

---

[67] *See* LAVANYA RAJAMANI, DIFFERENTIAL TREATMENT IN INTERNATIONAL ENVIRONMENTAL LAW 162 (2006) (CBDRRC "is a fundamental part of the conceptual apparatus of the climate change regime such that it forms the basis for the interpretation of existing obligations and the elaboration of future international legal obligations within the regime in question.").

[68] Article 3(1) of the UNFCCC states that: "The Parties should protect the climate system for the benefit of present and future generations of humankind, on the basis of equity". UNFCCC, *supra* note 1, art 3(1).

[69] UNFCCC, *supra* note 1, art 3(1). This usage is consistent with the history of international climate change negotiations with, for example, the UNGA Resolution on

The idea articulated here is referred to as the principle of intergenerational equity. The principle of intergeneration equity is commonly understood to "defin[e] the rights and obligations of present and future generations with respect to the use and enjoyment of natural and cultural resources, inherited by the present generation and to be passed on to future generations in no worse condition than received."[70] In common with CBDR, the principle of intergeneration equity is grounded in notions of distributive justice; and together with CBDR, it "articulates a principle of fairness among generations in the use and conservation of the environment and its natural resources".

The principle of intergeneration equity can be found expressed in numerous other places in international environmental law and is embedded in the principle of sustainable development[71], which is a driving normative force across the field of environmental law. Although interpretations vary and the principle lacks firm legal status in the corpus of international law, "the essential point of the theory, that [hu]mankind has a responsibility for the future, and that this is an inherent component of sustainable development, is incontrovertible, however expressed."[72] In common with CBDR, thus, the key question is how the principle will be realized through implementation of the climate change regime.

Concern for future generations is an inherent part of the climate change regime and can be found integrated in the UNFCCC, the Kyoto Protocol, COP decisions, and ongoing negotiations. Recognizing that greenhouse gases and in particular carbon dioxide may stay in the atmosphere for several hundreds of thousands of years, the treaty regime itself is premised on the need to address climate change "within a time-frame sufficient to allow ecosystems to adapt naturally to climate change, to ensure that food production is not threatened and to enable economic development to proceed in a sustainable manner."[73] This fundamental objective focuses on the

---

climate change that initiated intergovernmental negotiations. *See, e.g.,* UNGA Res. 45/212 *supra* note 6.

[70] Catherine Redgwell, *Principles and Emerging Norms in International Law: Intra and Inter-generational Equity,* in THE OXFORD HANDBOOK OF INTERNATIONAL CLIMATE CHANGE LAW (Cinnamon Carlarne, Kevin Gray, & Richard Tarasofsky eds.) (2016).

[71] The Brundtland Commission Report defines sustainable development as "development that meets the needs of the present without compromising the ability of future generations to meet their own needs" World Commission on Environment and Development, *Our Common Future* 43 (OUP, 1987). The UNFCCC, however, was one of the first treaties to place the principle of intergenerational equity firmly within the treaty of the text as a guiding principle for implementation.

[72] PATRICIA BIRNIE, ALAN BOYLE AND CATHERINE REDGWELL, INTERNATIONAL LAW AND THE ENVIRONMENT, 121 (3rd ed. 2009).

[73] UNFCCC, *supra* note 1, art 2(1).

long-term relationship between humans and the natural environment, taking into account the needs and well-being of future generations.

In practice, however, parties have struggled to implement the treaty, thus leaving both present and future generations at risk. These shortfalls are well recognized. In an effort to further effective implementation of the treaty in a way that ensures the well-being of present and future generations, the COP has institutionalized a goal since 2010 of limiting the increase in global temperature to no more than 2° Celsius above pre-industrial levels.[74] In 2015, under the auspices of the UNFCCC, a group of experts assessed the adequacy of the 2°C goal.[75] The report concluded that the 2°C goal is useful and still technically feasible, but that the world is not yet on track to meet this goal. The assessment further cautions that the 2°C goal should be "seen as an upper limit, a defence line that needs to be stringently defended, while less warming would be preferable."[76] For this reason, as noted above, the Paris Agreement sets 1.5° as an aspirational goal.

The 2°C goal serves as a vehicle for implementing the principle of intergenerational equity. The roadmap for implementing the 2°C goal, however, remains unwritten because, as discussed *infra,* even if fully implemented, existing NDCs fall short of allowing the global community to reach the 2°C goal. Similarly, while the Paris Agreement re-emphasizes the importance of adaptation and loss and damage mechanisms, it does little to create a defined and reliable pathway towards achieving comprehensive mitigation or adaptation goals.

C. The Precautionary Principle

The expert review of the 2°C goal recognized that the goal was achievable but stressed that "limiting global warming to below 2°C necessitates a radical transition in the form of deep decarbonization now and going forward, not merely a fine tuning of current trends."[77]

---

[74] The 2°C goal was first formally put forward in the 2009 Copenhagen Accord and was subsequently more institutionalized by the 2010 Cancun Agreements. *See* FCCC/CP/2010/7/Add.1, Decision 1/CP.16, para. 4 (recognizing "that deep cuts in global greenhouse gas emissions are required according to science, and as documented in the Fourth Assessment Report of the Intergovernmental Panel on Climate Change, with a view to reducing global greenhouse gas emissions so as to hold the increase in global average temperature below 2°C above pre-industrial levels, and that Parties should take urgent action to meet this long-term goal, consistent with science and on the basis of equity; also recognizes the need to consider, in the context of the first review . . . strengthening the long-term global goal on the basis of the best available scientific knowledge, including in relation to a global average temperature rise of 1.5°C").

[75] FCCC/SB/2015/INF.1, *Report on the Structured Dialogue of the 2013–2015 Review* (2015), http://unfccc.int/resource/docs/2015/sb/eng/inf01.pdf.

[76] *Id.* at 18 (Message 5).

[77] *Id.* at 11 (Message 2).

Acknowledging the challenges associated with meeting the 2°C goal, the expert review nevertheless advocates adopting a precautionary approach to addressing climate change that could require even more radical change. Specifically, the expert review suggests that "in the light of the difficulties in predicting the risks of climate change, there is value in taking a precautionary approach and adopting a more stringent target."[78]

This recent call for adopting a precautionary approach is rooted in the foundations of the climate regime. The UNFCCC calls for Parties to "take precautionary measures to anticipate, prevent or minimize the causes of climate change and mitigate its adverse effects" noting that "[w]here there are threats of serious or irreversible damage, lack of full scientific certainty should not be used as a reason for postponing such measures."[79]

While there is no definitive statement of the precautionary principle, the precautionary principle has been extensively incorporated since the mid-1980s, in varying iterations, in a wide variety of domestic and international environmental laws. The expression of the precautionary principle found in the UNFCCC mirrors the most commonly used statement of the principle, found in the Rio Declaration, which states: "[w]here there are threats of serious or irreversible damage, lack of full scientific certainty shall not be used as a reason for postponing cost effective measures to prevent environmental degradation."[80] Precautionary measures are particularly important when the failure to act rapidly in the present might lead to cumulative, irreversible, or potentially catastrophic negative effects down the road.

The problem of climate change lends itself particularly well to a precautionary approach. As Jonathan Wiener notes, "because climate change poses uncertain but serious and even catastrophic risks, and because greenhouse gas emissions have latent and long-lasting effects (over decades or centuries), precautionary action is widely urged as essential to preventing future climate change—rather than waiting to act after the damage is done and it is too late to address the cause."[81] Wiener further cautions that "it has long been understood that in the case of GHGs and climate change, 'a

---

[78] *Id.* at para 109. See also *id.* at p. 33 (Message 10) (Message 10 states: "Parties may wish to take a precautionary route by aiming for limiting global warming as far below 2°C as possible, reaffirming the notion of a defence line or even a buffer zone keeping warming well below 2°C.")

[79] UNFCCC, *supra* note 1, Art 3(3).

[80] *Rio Declaration on Environment and Development*, *supra* note 64, principle 15.

[81] Jonathan B. Wiener, *Precaution and Climate Change*, in THE OXFORD HANDBOOK OF INTERNATIONAL CLIMATE CHANGE LAW (Cinnamon Carlarne, Kevin R. Gray, & Richard Tarasofsky eds.) (2016).

wait-and-see policy may mean waiting until it is too late'."[82] In common with CBDR, however, there is widespread disagreement over the meaning and application of a precautionary approach to climate change.

In the early years of climate negotiations, conflict over adopting a precautionary approach was rife. Over time, as climate science has matured as a field, the scientific consensus on anthropogenic climate change has become nearly universal. As a result, the debate over the use of a precautionary approach to climate change has evolved from focusing on scientific uncertainty about the basic mechanics of climate change to focusing on more nuanced questions: what it means to adopt a precautionary approach to the massive problem of climate change and how to weight the potential costs and benefits of precautionary measures. Thus, even as scientific consensus on the anthropogenic forcing of the climate system has solidified, questions persist over the pros and cons of taking precautionary measures to limit greenhouse gas emissions. The primary upside of adopting a precautionary approach that includes sharp cuts in greenhouse gas emissions is limiting the possibility of a wide range of negative effects that range from potentially mild to potentially disastrous.[83] The challenges associated with adopting a precautionary approach, however, include potentially high short-term costs associated with hard-to-quantify long-term benefits as well as challenges associated with choosing which risk to focus on minimizing. As Wiener points out:

> Precautionary measures must select which risks to make top priority, and must confront their potential to affect multiple risks at the same time. Precautionary measures to prevent one risk may induce "side effects" or "risk-risk tradeoffs," including increases in other countervailing risks (ancillary harms), and decreases in other accompanying risks (ancillary- or co- benefits) . . .Yet many versions, or applications in practice, of the [precautionary principle] in regulatory policy tend to focus on only one salient risk at a time.[84]

---

[82] *Id.* (quoting: National Research Council, *CO2 and Climate: A Scientific Assessment* (1979)).

[83] The potential negative effects of climate change are well documented and include everything from threats to global food security and global biodiversity, to widespread displacement and resulting human rights and security challenges. The potential negative effects of climate change also include the possibility of disastrous change associated with "tipping points," that is, a threshold for abrupt and irreversible change associated with, *e.g.*, rapid decline in Arctic sea ice, Amazonian Forest die-back, monsoon disruption, or large-scale melting of land-based ice sheets.

[84] Wiener, *supra* note 81.

Notwithstanding the changing nature of the debate and the more nuanced analyses of taking varying precautionary approaches, "precautionary measures on climate change have been elusive."[85] So elusive, in fact, that by 2007, Lisa Heinzerling argued that "the precautionary moment for action on climate change—the period in which we might have acted based on something less than a scientific consensus on the causes and consequences of climate change—has passed. We are in a post-cautionary world now."[86]

The expert review of the 2°C goal and the ongoing round of climate negotiations reflect the idea that we are now in a post-cautionary world; a world where we already are committed to a certain degree of warming with all of the attending consequences, and the focus now is on keeping that warming from crossing a certain threshold that, if crossed, would result in costs that are deemed unacceptable.

Ongoing negotiations reflect a shift towards post-cautionary approaches across a number of issues beyond simply moving towards a damage-control framework for mitigation. These shifts include an increased focus on adaptation, as well as efforts to develop a liability mechanism (i.e., the Warsaw International Mechanism for Loss and Damage[87]) to cope with the inevitable effects of climate change. Similarly, outside of international climate negotiations, as discussed in Chapter 9, geoengineering has become a pervasive topic of conversation. Thus, even as the climate regime shifts towards developing post-cautionary approaches to limiting climate change and minimizing the negative effects of climate change, the precautionary approach remains relevant in a number of contexts, central among which is as a frame for thinking about the governance and use of geoengineering methods to limit the effects of greenhouse gases.

III. The Kyoto Protocol Mitigation Instruments and the Paris Agreement Article 6 Mechanisms

Even as negotiations have entered a post-cautionary phase, parties to the treaties continue to rely on a set of market-based mitigation tools introduced by the Kyoto Protocol. These instruments, commonly referred to as the Kyoto Protocol flexibility mechanisms, include: Emissions Trading (ET), the Clean

[85] *Id.*

[86] Lisa Heinzerling, *Climate Change, Human Health and the Post-Cautionary Principle*, O'Neil Institute for National and Global Health Law Scholarship, Research Paper No. 4, at 2 (2007).

[87] The Paris Agreement designates the Warsaw International Mechanism for Loss and Damage associated with Climate Change Impacts as the UNFCCC tool for addressing issues of loss and damage. Paris Agreement, *supra* note 3, Art 8(2).

Development Mechanism (CDM), and Joint Implementation (JI). Each of these three mechanisms is designed to help developed countries reduce greenhouse gas emissions in a cost-effective manner.

During treaty negotiations, the United States sought to include in the Protocol provisions allowing international trading in carbon emission allocations as a way to improve the cost-effectiveness of the climate regime.[88] The initial proposal was based, in part, on a successful, first-of-its kind emissions trading program that had been adopted as part of the 1990 Amendments to the U.S. Clean Air Act.[89] Initial proposals to introduce economic flexibility into greenhouse gas (GHG) mitigation measures received mixed response, with critics questioning both the morality of conferring a transferrable right to pollute and the practicality of devising such a complex mechanism during Protocol negotiations. In particular, there was a concern that the proposed measures would enable developed nations to invest primarily in emissions reductions efforts in poorer nations in order to avoid investing in more expensive emissions reduction measures at home. To some critics, this was simply a way to allow developed countries to "chea[t] on the basic commitment" to reduce domestic greenhouse gas emissions.[90] The European Union, in particular, was one of the most vocal critics of early proposals, fearing that emissions trading would allow the largest global emitters to shirk domestic emissions reductions responsibilities.[91]

In the end, the debate was settled in favor of including three types of market-based mitigation mechanisms—ET, JI, and CDM—in the Protocol without any concrete rules on how these instruments

---

[88] *See* Wiener *supra* note 81; MICHAEL GRUBB ET AL., THE KYOTO PROTOCOL: A GUIDE AND ASSESSMENT, 89–114 (1999).

[89] *See* Clean Air Act Amendments of 1990, Pub. L. No. 101-549, 104 Stat. 2399 (1990) (codified at 42 U.S.C. §§ 7401–7700 (1994)).

[90] Peter G. G. Davies, *Global Warming and the Kyoto Protocol*, 47 INT'L & COMP. L. QUARTERLY 446, 458 (1998).

[91] *See* ANDREW J. DESSLER & EDWARD A PARSON, THE SCIENCE AND POLITICS OF GLOBAL CLIMATE CHANGE: A GUIDE TO THE DEBATE 15 (2007). The European Union, paradoxically once one of the most vocal skeptics of economic flexibility mechanisms, went on to launch the world's largest emissions trading regime, the EU Emissions Trading Scheme (ETS), which is discussed in greater detail in Chapter 4. Moreover, EU member states have been responsible for a majority of the verified CDM and JI projects. European Union climate policy now fully embeds the flexibility mechanisms in regional efforts to reduce greenhouse gas emissions. *See* Directive 2004/101/EC of the European Parliament and of the Council of 27 October 2004 amending Directive 2003/87/EC establishing a scheme for greenhouse gas emission allowance trading within the Community, in respect of the Kyoto Protocol's project mechanism.

should be complemented by domestic emissions reduction efforts.[92] These mechanisms are discussed in detail below.

Over time, the flexibility mechanisms have become the Kyoto Protocol's greatest legacy. Each of the three programs has evolved over time and provided fertile ground for experimentation in different types of mitigation activities. Although each of the mechanisms is introduced here, Chapter 4 offers further insight into the use of different mitigation mechanisms in practice.

A. The Clean Development Mechanism (CDM)

The CDM[93] is designed to stimulate economically efficient greenhouse gas emissions reductions, while concurrently promoting sustainable development in developing countries.[94] It is the only of the three Kyoto flexibility mechanisms that allows developed and developing countries to cooperate on emissions reduction projects for mutual benefits. The CDM is a project-based mechanism. The objective of the program is to encourage investment in emissions reductions projects in developing countries.[95] Successful projects produce economic benefits for the host country (i.e., developing country) and certified emission reductions (CER) units that developed country parties can use to comply with their emissions reduction commitments under the Kyoto Protocol. Each CER generated is equivalent to one metric ton of CO_2. CERs are fungible commodities that can be sold and traded, including within existing emissions trading systems such as the EU ETS.

The CDM has been heralded as one of the most successful components of the Kyoto Protocol. There are currently in excess of 7803 registered projects and 313 registered programs of activities that are expected to generate almost 2 billion tons of carbon-dioxide-

---

[92] For example, Article 12, which pertains to emissions trading, merely states that: "Any such trading shall be supplemental to domestic actions for the purpose of meeting quantified emission limitation and reduction commitments under that Article."

[93] Kyoto Protocol, *supra* note 2, art 12.

[94] Sustainable development criteria are left to each member state to define, but combine elements of economic development, social development, and environmental protection. These three components are often referred to as the three pillars of sustainable development. *See, e.g.,* 2005 World Summit Outcome, G.A. Res. 60/1, P 48, U.N. Doc. A/RES/60/1 (Oct. 24, 2005).

[95] Over time, the CDM has expanded to include both projects, as originally understood to constitute individual, one-off investments, as well as programmes of activities (PoA), which make it possible "to register the coordinated implementation of a policy, measure or goal that leads to emission reduction. Once a PoA is registered, an unlimited number of component project activities (CPAs) can be added without undergoing the complete CDM project cycle." Expanding CDM to include PoA creates greater opportunities for investment in less developed countries where there are fewer opportunities for free-standing, large-scale emissions reductions projects.

equivalent reductions.[96] In addition to generating billions of tons of reductions in greenhouse gas emissions, the CDM also generates millions of dollars in adaptation funding for developing countries. This is because the Kyoto Protocol directs the Parties to this Protocol to "ensure that a share of the proceeds from certified project activities is used to . . . assist developing country Parties that are particularly vulnerable to the adverse effects of climate change to meet the costs of adaptation."[97] This mandate was realized through the creation of the Adaptation Fund by the Parties to the Protocol in 2001 as a mechanism for financing concrete adaptation projects and programs in the developing world.[98] The Adaptation Fund is financed, in part, by a levy on CDM projects that amounts to 2 percent of CERs generated by each project.

Despite its apparent success in generating significant emissions reductions and adaptation resources, the CDM has been beset by administrative, effectiveness, and equity concerns. These concerns include fears that drawing upon CDM credits, at the expense of implementing domestic abatement strategies, impedes efforts to reduce domestic greenhouse gas emissions and results in lost opportunities to update infrastructure and operations in the energy sector and to capture economic opportunities in the global green-energy market. Parallel criticisms focus on the fact that the CDM allows the developed world to capture all of the developing world's low-hanging fruit; that is, the argument goes, CDM projects capture the easiest and most profitable opportunities to reduce greenhouse gas emissions thus making it more difficult and expensive for the host country to achieve additional emissions reductions down the road.

Beyond concerns about displacing domestic action and driving up future costs of compliance, the CDM board[99] has struggled to devise effective implementation and accounting rules. As Shi-Ling Hsu explains:

> [i]n the early days of implementation, the CDM . . . produced numerous mistakes of overinclusion—approval of projects that did not really reduce emissions—and of

---

96 UNFCCC, *Achievements of the Clean Development Mechanism: Harnessing Incentive for Climate Action, 2001–2018*, https://unfccc.int/sites/default/files/resource/UNFCCC_CDM_report_2018.pdf.

97 Kyoto Protocol, *supra* note 2, art 12(8).

98 *See* Report of the Conference of the Parties on its Seventh Session, held at Marrakesh from 21 October to 10 November 2001, U.N. Doc. FCCC/2001/13, Add.1, Decision 10/CP.7 (2002), https://unfccc.int/documents/2516.

99 The supervisory board for the CDM is the CDM Executive Board (EB). The CDM EB operates "under the authority and guidance of the Conference of the Parties serving as the Meeting of the Parties to the Kyoto Protocol (CMP)." CDM, Governance, http://cdm.unfccc.int/EB/governance.html.

underinclusion—the rejection of projects that would have produced an emissions reduction. For instance, a huge number of CDM projects in China purported to reduce emissions of HFC-23, a powerful GHG and byproduct of the production process, generating credits that could be used by emitters in developed countries in lieu of actually reducing GHG emissions. The problem was that the value of the credits far exceeded the value of the captured refrigerants. The plants producing HFC-23 had no real purpose other than the generation of credits; refrigerants were a mere pretense for such generation. The issuance of these credits subjected the CDM Board to considerable criticism.[100]

In light of these and similar problems, the CDM Board has continually reviewed and modified the rules and modalities in an effort to develop better strategies for, among other things: determining what types of projects are valid; establishing emissions baselines and monitoring regimes; ensuring that the aimed for emissions reductions would not have been achieved absent investment in the CDM project; and developing uniform tools for certifying emissions reductions from a wide range of projects.

Alongside the effectiveness questions, equity concerns have plagued the development of the CDM. The CDM has been criticized on equity grounds, including claims that the projects are not promoting sustainable development or technology transfer in the host countries and that the projects are disproportionally located in middle income economies to the detriment of the lower income economies. The CDM has also come under fire for creating burdensome bureaucratic processes that make it difficult for many developing countries to participate.

Through a continual process of institutional evaluation, the Parties to the Protocol have sought to improve the effectiveness and fairness of the CDM. While many questions persist, the CDM is the largest international emissions offsetting scheme, thus far.

B. Joint Implementation (JI)

The second largest offsetting scheme after the CDM is JI. In common with CDM, JI is a project-based flexibility mechanism. Unlike CDM, participation in JI projects is limited to UNFCCC Annex I parties, including developed country parties and the economies in transition.[101] Under JI, any Annex I party is eligible to

---

[100] Shi-Ling Hsu, *International Market Mechanisms: An Introduction*, in THE OXFORD HANDBOOK OF INTERNATIONAL CLIMATE CHANGE LAW (Cinnamon Carlarne, Kevin R. Gray, & Richard Tarasofsky eds.) (2016).

[101] The economies in transition (EIT) include the Russian Federation and several other Central and Eastern European Countries, *e.g.*, Russian Federation,

participate in the buying and selling of credits, known as emission reduction units (ERU), that are generated through a JI approved project in an Annex I host country.[102]

The CDM began and grew quickly, with the first project being registered on the same day that the Kyoto Protocol came into force in 2005. In contrast, the JI program was slower to get off the ground. The Parties to the Protocol did not finalize guidelines for JI until 2005[103], with the first projects coming on-line in 2008.

There are two categories of JI projects: 'Track 1' and 'Track 2'. 'Track 1' projects, otherwise known as 'Party-verified' projects, provide a simplified project approval process. 'Track 1' projects are verified by the host Party without requiring external verification by the JI Supervisory Committee. In order to use 'Track 1', a host Party has to fulfill a set of eligibility requirements that demonstrates that it is has adequate procedures in place to verify emissions reductions from the project. If the host Party has satisfied these requirements, it will be able to issue ERUs for JI projects. 'Track 1' reduces the amount of bureaucratic red tape associated with project approval and verification and is the preferred method for project approval, accounting for approximately 97 percent of all approved projects. If a host Party is unable to meet the eligibility requirements, the project will follow the procedures for verification under 'Track 2'. For 'Track 2' projects, ERUs must be verified by the JI Supervisory Committee.[104]

While JI was slow to get off the ground, between 2008–2012, the number of projects expanded. There are now 17 Parties hosting 'Track 1' projects. The economies in transition, including the Russian Federation, Ukraine, and the Czech Republic, host a majority of the projects, but other Annex I Parties such as New Zealand, Sweden, France, and Germany also host projects. JI projects include a range of activities, including investments in wind farms, biomass, biogas, combined heat and power, fugitive emissions capture.

Estonia, Belarus, Ukraine, Latvia, Lithuania, Hungary, Bulgaria, Czech Republic. These countries, while being treated as developed countries for purposes of the climate change regime, are subject to fewer obligation in certain respects, including the provision of funding and technology transfer.

[102] Article 6 provides: "any Party included in Annex I may transfer to, or acquire from, any other such Party emission reduction units resulting from projects aimed at reducing anthropogenic emissions by sources or enhancing anthropogenic removals by sinks of greenhouse gases in any sector of the economy." Kyoto Protocol, *supra* note 2, art 6(1).

[103] *Implementation of Article 6 of the Kyoto Protocol,* Decision 10/CMP .1, FCCC/KP/CMP/2005/8/Add.2 (March 30, 2006), http://unfccc.int/resource/docs/2005/cmp1/eng/08a02.pdf#page=2.

[104] The JI Supervisory Committee is an international body constituted under the authority and guidance of the Parties to the Protocol.

In common with CDM, JI struggles with problems of effectiveness including questions surrounding setting current baselines, ensuring projects are achieving additional emissions, and reviewing ERU validation processes. Similarly, the future of both CDM and JI hangs in the balance leading up to Paris.

C. Emissions Trading

CDM and JI create project-based mechanisms for achieving emissions reductions. In contrast, the third flexibility mechanism, Emissions Trading (ET), facilitates the creation of a market in tradable emissions reductions credits. Emissions trading will be discussed in detail in Chapter 4. For purposes of introduction, the basic idea behind emissions trading is that a regulatory entity confers a property right to an environmental resource and then creates a market for trading in that property right.

The Kyoto Protocol embraces ET in Article 17, allowing the developed countries to "participate in emissions trading for the purposes of fulfilling their" emissions reductions commitments and calling on the Conference of the Parties to "define the relevant principles, modalities, rules and guidelines, in particular for verification, reporting and accountability for emissions trading."[105]

A decade after the Kyoto Protocol came into force, ET schemes have proliferated. Most schemes are domestic or regional in scope, the largest of which is the EU Emissions Trading Scheme.[106] Yet, there is still no global carbon market of the kind many envisioned in the wake of the Protocol negotiations and many of the existing trading programs are still struggling to create healthy, well-functioning markets.

Emissions trading schemes create opportunities to maximize the cost-effectiveness of climate mitigation measures, but they also pose accountability, transparency and legitimacy challenges that create opportunities for abuse and market failures.[107] In addition, in common with the other flexibility mechanisms, ET creates a series of under-explored equity questions. As a general matter, integrating flexibility into climate policies skews assessments of distributive justice by placing economic rationality at the center of decision-

---

[105] Kyoto Protocol, *supra* note 2, art. 17. Here, again, the Protocol dictates that ET should be supplemental to domestic emissions reductions but does not specify to what degree, stating that: "[a]ny such trading shall be supplemental to domestic actions for the purpose of meeting quantified emission limitation and reduction commitments under that Article." *Id.*

[106] The EU Emissions Trading Scheme (ETS) is examined in greater detail in Chapter 4.

[107] Institutional misuse, both intentional and unintentional, almost undermined the first phase of the EU ETS, for example.

making processes without fully assessing the implications for the distribution of privileges and obligations.

Despite existing challenges, emissions trading continues to be a favorite strategy for reducing greenhouse gas emissions across the globe.

D. The Paris Agreement Article 6 Mechanisms

Like the Kyoto Protocol, the Paris Agreement embraces the idea that market-based mechanisms, or what the Agreement refers to as "cooperative approaches" in Article 6, are a fundamental part of a low emissions development pathway. This embrace of cooperative mechanisms builds upon the flexibility mechanisms created by the Kyoto Protocol—including the Clean Development Mechanism, Joint Implementation, and Emissions Trading—but differs from these previous tools in significant ways. Paralleling the cooperative and inclusive model embodied by the NDC approach to mitigation, the Paris Agreement designs the cooperative approaches to be fully open to all state parties. This approach expands the scope of potential carbon markets and mitigation cooperation and creates a more open field than under the Kyoto Protocol, where the only avenue for developing state participation was under the CDM, where their role was limited to hosting mitigation projects.

As envisioned by the Paris Agreement, the Article 6 mechanisms are broad and allow parties to cooperate, using market and non-market tools, to address climate change. Article 6(1) begins by recognizing that "some Parties choose to pursue voluntary cooperation in the implementation of their nationally determined contributions to allow for higher ambition in their mitigation and adaptation actions and to promote sustainable development and environmental integrity."[108] It then sets out what appear to be parameters for tradable units, a cooperative mechanism similar to the Clean Development Mechanism, as well as a framework for non-market cooperative approaches. In key part, Article 6.2 provides bilateral actions to reduce or remove GHG emissions; Article 6.4 creates a new multilateral mechanism designed to improve upon and replace the CDM; and Article 6.8 provides for non-market cooperation among governments.

Focusing here, on the creation of a bilateral tradable unit and a new multilateral mitigation mechanism, Article 6(2) provides in relevant part that:

> Parties shall, where engaging on a voluntary basis in cooperative approaches that involve the use of

---

[108] Paris Agreement, *supra* note 3, Art 6(1).

> internationally transferred mitigation outcomes towards nationally determined contributions, promote sustainable development and ensure environmental integrity and transparency, including in governance, and shall apply robust accounting to ensure, inter alia, the avoidance of double counting, consistent with guidance adopted by the Conference of the Parties serving as the meeting of the Parties to this Agreement.[109]

This provision provides for the creation of an "internationally transferred mitigation outcome" (ITMO); ITMOs are envisioned as a form of tradable credit. Article 6(4) then provides for a second form of flexible mitigation mechanism when it establishes a "mechanism to contribute to the mitigation of greenhouse gas emissions and support sustainable development".[110] The objectives of this new mechanism, which colloquially was referred to early on as the Sustainable Development Mechanism (SDM) are to: promote the mitigation of greenhouse gas emissions while fostering sustainable development; incentivize and facilitate participation in the mitigation of greenhouse gas emissions by public and private entities; reduce emission levels in the host Party while also allowing another Party to fulfill its NDC; and deliver an overall mitigation in global emissions.

Together, these two provisions of Article 6 were generally understood to create first, units that could be traded or otherwise transferred among countries, potentially allowing for the creation of an international trading mechanism, and second, the framework for a new cooperation mechanism that would mirror but replace the CDM.

Between 2015–2021, the Parties to the Agreement worked to develop the modalities for the Article 6 mechanisms through the Paris rulebook. These efforts culminated in 2021 at COP26 in Glasgow, Scotland, where the Parties successfully developed the basic framework for the new cooperative mechanisms.

In Glasgow, the Parties to the Agreement finalized three key decisions related to the implementation of the Article 6 provisions.[111]

---

[109] Paris Agreement, *supra* note 3, Art 6(2). Art 6(3) then states that: "The use of internationally transferred mitigation outcomes to achieve nationally determined contributions under this Agreement shall be voluntary and authorized by participating Parties."

[110] *Id.* Art 6(4).

[111] UNFCCC, Decision 2/CMA.3, *Guidance on Cooperative Approaches Referred to in Article 6, Paragraph 2, of the Paris Agreement*, FCCC/PA/CMA/2021/10/Add.1 (March 8, 2022), https://unfccc.int/sites/default/files/resource/cma3_auv_12a_PA_6.2.pdf; UNFCCC, Decision 3/CMA.3, *Rules, Modalities and Procedures for the Mechanism Established by Article 6, Paragraph 4, of the Paris Agreement*, FCCC/PA/CMA/2021/10/Add.1 (March 8, 2022), https://unfccc.int/sites/default/files/resource/cma3_auv_12a

These decisions focus on ensuring that all cooperative approaches create real, additional, and verifiable emissions reductions and on creating common metrics to ensure "transparency, accuracy, completeness, comparability, and consistency" in how emissions reductions are calculated.[112] In key part, the new rules for the Article 6 mechanisms center on ensuring that GHG emission reductions cannot be double counted (i.e., counted by both the host and home country), limiting the number of past CDM projects that countries can count towards their reporting under their NDCs, establishing a new (unnamed) international mechanism to approve projects under Article 6(4); and designating that an equivalent of 5% of the "share of proceeds" generated pursuant to the new 6(4) multilateral mechanism will be transferred to the Global Adaptation Fund to aid in adaptation efforts in developing countries.[113] Notably, the new rules apply to both government-to-government transactions and government-to-private sector transactions.

The developments in Glasgow reaffirm the Parties' commitment to using market-based strategies as an important tool for limiting greenhouse gas emissions and achieving the goals of the Paris Agreement.

IV. The Future of the International Climate Regime

The international climate regime is complex and evolving. The UNFCCC, as the centerpiece for this regime, defines the operational and normative parameters for global discourse and provides an essential forum for dialogue and decision-making. It is, and always has been, the focal point for the development of innovative instruments and principles of international environmental law in the climate change context. The parties to the UNFCCC, however, confront a near-impossible challenge. They are tasked with framing a global approach to addressing climate change that informs state actors as to their legal and moral obligations.[114] This international

_PA_6.2.pdf; UNFCCC, Decision 4/CMA.3, *Work Programme Under the Framework for Non-Market Approaches Referred to in Article 6, Paragraph 8, of the Paris Agreement,* FCCC/PA/CMA/2021/10/Add.1 (March 8, 2022), https://unfccc.int/sites/default/files/resource/cma3_auv_12a_PA_6.2.pdf.

[112] *Guidance on Cooperative Approaches Referred to in Article 6, Paragraph 2, of the Paris Agreement, supra* note 111, at III(B).

[113] Charles E. Di Leva & Scott Vaughan, *The Paris Agreement's New Article 6 Rules,* International Institute for Sustainable Development (Dec. 13, 2021), https://www.iisd.org/articles/paris-agreement-article-6-rules.

[114] Or, legal outcomes, as the case may be. While the 1990 U.N. General Assembly resolution that launched the U.N. climate change negotiations and, ultimately, culminated in the adoption of the UNFCCC called for "a framework *convention* on climate change, and other related instruments, containing appropriate *commitments* for action to combat climate change and its adverse effects," the more recent Durban Platform calls for a more generalized process to develop a "protocol,

framework, in turn, is envisioned as incentivizing state-based efforts to address climate change that, collectively, address a massive problem. The daunting nature of this task has become increasingly apparent, as the international community has struggled to develop the contours of an effective regime. Over time, it has become clear that the international climate regime does not, and cannot offer a simple linear, top-down solution to the global challenge of climate change.

As demonstrated in the detailed discussions throughout this book, after over thirty years of global negotiations, our understanding of the massive nature of the problem has matured.[115] In the lead-up to the Paris COP, the global community found itself at the precipice of creating a new legal strategy for addressing climate change. The Paris Agreement upends the Kyoto Protocol-style, top-down approach to addressing climate change and it creates greater room for communication, cooperation, and experimentation with respect to both mitigation and adaptation. Moreover, the Glasgow Climate Pact represents strengthening international resolve to achieve the goals of the Paris Agreement. Despite this progress, the evolving international climate change regime still falls short of creating a robust pathway towards preventing dangerous climate change. Future efforts to address global climate change must acknowledge the limitations of the international regime and focus on mobilizing international, domestic, and local resources, from both the public and private sector to structure more effective mitigation and adaptation strategies.

Further Readings

Lisa Benjamin & David Wirth, *From Marrakesh to Glasgow: Looking Backward to Move Forward on Emissions Trading*, 3–4 CLIMATE LAW 245 (2021).

---

another legal instrument or a legal outcome," and it entirely omits any language about commitments. UNGA Res. 45/212, *supra* note 6; *Establishment of an Ad Hoc Working Group on the Durban Platform for Enhanced Action*, UNFCCC Dec. 1/CP.17, para. 2 (Dec. 11, 2011). In this way, the Durban Platform creates a mandate for some manner of legal agreement, but the mandate is imprecise; the meaning of "legal outcome" is sufficiently ambiguous to offer an escape hatch if, in three years' time, the parties to the UNFCCC remain divided. As a result, the future of climate change mitigation efforts remains uncertain at best. Cinnamon Carlarne, *Rethinking a Failing Framework: Adaptation and the Future of the Global Climate Change Regime*, 25 GEO. INT'L ENVTL. L. REV. 1, 18 (2012).

[115] *See, e.g.*, J.B. Ruhl & James Salzman, *Climate Change, Dead Zones, and Massive Problems in the Administrative State: A Guide for Whittling Away*, 98 CALIF. L. REV. 59, 72–80 (2010) (suggesting a framework for looking at complex problems whereby the problem is approached as a "massive problem" that requires more precise definition and is capable of being more effectively—if not fully—addressed through carefully crafted regulatory frameworks).

Daniel Bodansky, *The Paris Climate Change Agreement: A New Hope?*, 110 AM. J. INT'L L. 288 (2016).

Cinnamon P. Carlarne & JD Colavecchio, *Balancing Equity and Effectiveness: The Paris Agreement and the Future of International Climate Change Law*, 27 N.Y.U. ENV'T L.J. 107 (2019).

Cinnamon Carlarne, *Delinking International Environmental Law and Climate Change,* 4 MICH. J. ENVTL. & ADMIN. L. 1 (2014).

Lavanya Rajamani & Daniel Bodansky, *The Paris Rulebook: Balancing International Prescriptiveness with National Discretion*, 68 INT'L & COMPARATIVE L. QUARTERLY 1023 (2019).

Lavanya Rajamani, *Ambition and Differentiation in the 2015 Paris Agreement: Interpretative Possibilities and Underlying Politics*, 65 INT'L & COMPARATIVE L. QUARTERLY 493 (2016).

Harro van Asselt, *The Design and Implementation of Greenhouse Gas Emissions Trading*, in CINNAMON CARLARNE, KEVIN GRAY, & RICHARD TARASOFSKY eds., THE OXFORD HANDBOOK OF INTERNATIONAL CLIMATE CHANGE LAW (2016).

Michael Wara, *Measuring the Clean Development Mechanism's Performance and Potential*, 55 UCLA LAW REVIEW 1759 (2008).

Chapter 4

PUTTING A PRICE ON CARBON

In later chapters, we will look at what states and the federal government are doing to reduce carbon emissions. First, however, we need to examine the range of instruments that are available for this purpose. Economists particularly favor the use of carbon markets and carbon taxes as opposed to conventional methods of pollution control. There is now significant experience around the world with these tools. Just to be clear on terminology, we will be considering three possible approaches to reducing carbon in this chapter:

Conventional Regulation. Sometimes pejoratively called command-and-control regulation. This type of regulation may involve mandates to use a specific technology but usually takes the form of a performance requirement. For example, a utility might be told to reduce its carbon emissions by a specified percentage over a specific period of time. Such regulations are often based on economic feasibility, but sometimes attempt to force the development of new technology by setting a standard beyond the current capability of industry. There is no explicit price tag on carbon, but the cost of compliance per ton of emission reduction sets an implicit carbon price.

Emissions Trading. Also called cap and trade. The government sets a cap on emissions from some set of sources and allocates or auctions permits (also called allowances), generally allowing one ton of emissions per permit. The permits total the cap. The distinctive feature is that firms can buy and sell permits. In effect, the market price for permits is the price of carbon. But this may not be known until the scheme is already in operation.

Emissions Tax. The government imposes a tax on carbon emissions, which might be collected from emitters or might be collected "upstream" from sellers of fossil fuels. The revenue can be used to offset reductions in other taxes or can be spent on environmental or other purposes. The price of carbon is simply the amount of the tax. In principle, this should be set to equal the social cost of carbon, so that emitters fully internalize the harms they cause. As we have seen, however, there is considerable uncertainty about the social cost of carbon.

This chapter will begin with a general discussion of how the various tools work and of their strengths and drawbacks. There is a robust body of literature by economists and others debating the merits of these different approaches. Perhaps it is not surprising that environmentalists and economists have different views on this subject. For that reason, the debate has sometimes been a heated one.

We will then look at experiences with using market tools to date. Both carbon trading and carbon taxes have been implemented in a number of jurisdictions, though trading seems to have gotten significantly more use. Given this experience, we may be able to make a more informed choice of instruments. Instrument choice involves competing considerations, not the least of which are political. In general, however, we are interested in achieving climate goals as effectively and inexpensively as possible. Climate change mitigation is a huge undertaking, and we do not want to unnecessarily waste resources if we can achieve our goals more cheaply.

Carbon mitigation efforts are generally aimed at energy-related emissions. However, emissions restrictions can be combined with a system of credits based on reductions in non-energy emissions. For instance, a regulation might require a utility to reduce its carbon by a certain amount, but give the company credit toward that reduction if it has taken steps to reduce emissions from agriculture or deforestation. (Imagine, for example, an agreement by a power plant to plant trees in order to remove carbon from the atmosphere.) Thus, the reduced emissions from the other sources could "offset" the firm's own emissions. Such offsets can also be used in trading schemes, and tax credits can serve a similar purpose under a carbon tax. The third section of the chapter will consider both the potential benefits and problems of these offsets.

The fourth section considers a problem common to all of these schemes: unless the carbon restriction covers the entire world and all types of emissions, efforts to reduce carbon emissions in one way or at one location can result in higher emissions somewhere else. In effect, the carbon "leaks" from the regulated activity, reducing the benefits of regulation. Regulators need to consider the possible seriousness of this problem and potential policy responses. Obviously, a jurisdiction wants to avoid a situation where it works hard to reduce carbon emissions but just as much carbon ends up going into the atmosphere from somewhere else, cancelling out its efforts.

From a policy perspective, the choice of regulatory instruments is critical. Even a lawyer whose interest is limited to counseling clients needs to have an understanding of the policy issues. Climate regimes are still in flux, so the debates in this area have real practical import. Lawyers will be involved in the processes of legislation and administration rulemakings where these decisions take place.

While our focus is on government use of carbon pricing as a policy instrument, many private firms have also begun placing a price on carbon for planning purposes. In 2021, McKinsey reported that "23 percent of the approximately 2,600 companies in our data set indicated they are using an internal carbon charge, and another 22 percent plan to do so in the next two years."[1] Some firms go further. One major American bank applies a carbon price to the operations of borrowers in considering credit risks. A leading software company charges its business groups a small carbon fee and uses the funds to support internal efficiency initiatives, green power, and carbon offset projects. If nothing else, these business actions reflect an expectation that emitting carbon will become increasingly costly in the future.

I. Trading, Taxes, and Regulation

Economists offer market instruments (trading and taxes) as an alternative to more conventional types of emissions regulation. Conventional regulations are familiar from other environmental contexts, such as pollution control requirements for factories. These regulations often require industry to use the "best available technology" (BAT). Sometimes, however, they are set at a level based solely on the risk of the activity to public health, as with the Clean Air Act's air quality standards. Economists, however, tend to be skeptical of conventional regulation because they view it as unable to tailor emission limits to the costs and benefits of controlling individual sources.

One major unknown in choosing a regulatory approach relates to innovation. To reach longer-term climate goals, new technologies will have to be deployed. We do not have strong evidence about what approach is best for stimulating the development and uptake of new technologies. Clearly, this will not happen under any approach if emitting carbon is relatively cheap and unrestricted. But whether a tax, a trading system, or conventional regulation does the most to stimulate innovation is unclear. And any of these systems might need to be supplemented with other incentives, such as government funding for energy research or large prizes for technological breakthroughs. In this chapter, we will limit ourselves to considering how these various systems might work under relatively short time-frames like a decade, where these dynamic considerations are less of a concern.

---

[1] Jessica Fan, Werner Rehm, and Giulia Siccardo, *The State of Internal Carbon Pricing* (February 10, 2021).

A. Conventional Regulation Versus Emissions Trading

To understand the arguments about traditional regulation versus market instruments, it may be helpful to start with a simplified example of traditional regulation. Suppose a state wants to increase the use of solar energy so as to limit carbon emissions. Consequently, the legislature instructs the state public utility commission (PUC) to require "the greatest practicable use" of solar. This is the kind of formulation that is used in many pollution laws. To carry out this law, the PUC would need to obtain engineering and economic information about utilities in order to decide how much use of solar would be practical for a given utility. It could then issue mandates telling each utility how much of its electricity would have to come from solar.

There are some obvious advantages to this approach. Although the PUC will have to make some judgment calls, most of its task involves purely technical judgments about economic costs and engineering potential. Once the PUC sets the standard for each utility, the utility's task is clearly defined. Moreover, enforcement is relatively easy, since it is merely necessary to check how much electricity is being purchased by the utility from solar sources and how much it generates itself.

But such a law would also have some problems. First, the PUC is likely to have less expertise and data about utilities than the utilities themselves. The utilities have every incentive to exaggerate the cost and engineering difficulties of expanding the use of solar. Thus, the PUC is at an informational disadvantage. The regulated parties have the most information but no incentive to act on it.

Second, by limiting its mandate to solar, the legislature may have focused too much on a single technology. Although solar may look promising, perhaps in the end wind power will turn out to be a more practical energy source, at least for some utilities. Or perhaps it will be easier for utilities to cut their emissions by encouraging energy efficiency on the part of their customers. A solar-only mandate may turn out to be more expensive and difficult than a more flexible mandate. For that reason, many states have adopted renewable portfolio standards, requiring that utilities obtain a designated percentage of electricity from a specified group of renewables, but leaving them flexibility within this group.

Imposing a uniform floor on renewables for all utilities raises another problem. Some utilities might find it much easier to incorporate renewables into their electricity mix than others. This might be true because of local weather conditions or other physical constraints or perhaps because their alternate energy sources are already expensive (so the price of buying renewables is closer to their existing costs). The state could obtain the same uptake of renewables at lower costs if it allowed more of the burden of renewable use to fall on the firms where it posed the least burden. One way of doing that is to allow firms to buy or sell renewable credits. For instance, a firm that needed 100 megawatts of renewables to meet its quota, but actually had 110 megawatts, could sell 10 megawatts in renewable credits to another firm that had only 90 megawatts of capacity. That way, both firms would satisfy their obligations; the firm with the most renewables would be rewarded while the laggard firm would be out of pocket for the prices of the credits. These marketable renewable credits are a common feature of renewable portfolio standards or RPSs. This approach presents a double incentive to develop renewable energy, since it either reduces a cost to the firm in having to buy credits, or it creates a profit motive by allowing the firm to sell credits.

Trying to achieve a cost effective allocation of carbon reduction is a powerful idea. RPSs often have restrictions, such as how much electricity can come from specific renewables or limits on the definitions of renewables, that compromise their ability to reach this goal. Moreover, reducing carbon emission from utilities might itself not be the most cost effective approach. After all, reducing carbon emissions from other sources such as industrial boilers might be cheaper. The market will be most efficient if it is as broad as possible, so that there is an incentive to take advantage of all possible opportunities to reduce carbon in a cost-effective way. In response to this consideration, we could imagine expanding the RPS requirement to include all carbon sources, transferring responsibility from the PUC to the state's environmental regulatory agency in order to include a broader range of sources. At that point, we would find ourselves with a full-scale carbon trading scheme.

A cap-and-trade scheme is essentially akin to a "universal RPS." It puts a limit on total carbon emissions. It then issues allowances that permit the emission of one ton of carbon, with the total number of allowances being set in advance. This is the *cap*. At the end of the year, each emitter must hold a sufficient amount of allowances to cover all its emissions that year. The allowances can be traded, so an emitter that has more emissions than its original allotment can buy more allowances from other firms. Some emitters may find it feasible to cut their emissions cheaply so they do not need all of their allowances, which they can then sell. In general, if a firm's emission-reduction costs are above average, it can benefit from buying allowances. Similarly, a firm with below-average emission-reduction costs can benefit from selling allowances. In a well-functioning market, the result should be to equalize the cost of emissions reductions across all firms. Logically, this has to be the cost-efficient way of reaching the cap, since no firm can reduce a ton of emissions any more cheaply—otherwise, it would have taken advantage of this opportunity to sell extra allowances, driving down the market price for allowances.

One important design question is how to allocate allowances. Allowances may be distributed for free, often on the basis of past emissions (grandfathering). For instance, each firm might receive allowances equal to some percentage of its average emissions over the five years before the scheme is adopted. Grandfathering can raise equity and incentive concerns. Firms that took the lead in reducing their emissions before the plan is adopted will be penalized with lower allocations of emissions, while laggards will get more allocations.

Economists are critical of free allocation of allowances. The reason is that if a cap is effective in reducing emissions, it almost inevitably raises prices. If companies get allowances for free but are allowed to profit from higher prices, they benefit at the expense of consumers. Instead, economists favor auctioning allowances, so that the higher revenue received by companies becomes available for public use. From an equity perspective, it can also be argued that free allocation allows companies to make free use of the atmosphere, a public resource, for carbon disposal. For political reasons, however, it seems common to use free allocation, at least in the early years of a trading scheme, before possibly shifting to auctioning. Companies naturally favor free allocation, and consumers may be under the misimpression that making allowances free will limit price increases.

Emissions markets, at least in theory, offer a mechanism for achieving environmental goals such as carbon reductions at the lowest price possible. It is hard to argue against the desirability of minimizing costs while still attaining environmental targets. But emissions markets also have their critics. One criticism is that they send the wrong message by creating property rights in emissions in the form of allowances. Critics believe that doing so legitimizes pollution.

A second criticism is that emissions markets can be complicated and difficult to design. The design difficulties can create unforeseen snags or opportunities to avoid compliance. As the financial meltdown of 2009 illustrates, even markets that appear very sophisticated and well-functioning can fail spectacularly on occasion, and this is a risk that worries some opponents of emissions trading.

Outside of the climate change arena, a vivid example of the possible pitfalls of emissions trading is provided by RECLAIM, Southern California's nitrogen oxides (NO_X) and SO_2 trading program.[2] RECLAIM has produced a mixed record. Initial over-allocation of permits, which seems to be a common problem in trading systems, provided no real incentive to install control technologies. In other words, the cap was ineffective because it was set above the level that firms would have emitted anyway. After this problem was addressed, the trading system broke down for another reason: the California electricity crisis in 2009, which ultimately forced the removal of the power sector from the NO_X market. Essentially, when electricity wholesale prices spiked in 2009 due to a mishandled deregulatory effort, firms ramped up use of generators that lacked pollution controls, using the emissions credit market to offset the increased pollution. Prices for emissions credits shot through the roof, discrediting the RECLAIM scheme and leading to major reforms.

As this experience illustrates, emissions trading systems often face initial problems and need revamping before they operate smoothly. One lesson learned from the past quarter century of experience with emissions markets is that they are a good deal more difficult to structure effectively than the theory would suggest. In the case of RECLAIM, the midcourse correction seems to have worked.

A third criticism of emissions trading comes from advocates of environmental justice, who are concerned because it is hard to predict in advance what firms will end up reducing emissions and which will simply buy allowances (perhaps even expanding their emissions in the process). The result could be local hotspots, or areas where a high number of emitters are co-located. This possibility is not very significant in terms of climate change since CO_2 is well mixed in the atmosphere. But efforts to reduce CO_2 emissions will generally have the side-benefit of reducing other emissions, which may have more localized impacts. If the emissions market results in a geographic concentration of firms buying up carbon allowances and emitting more carbon, their emissions of conventional pollutants will also be higher. Advocates fear that these uncontrolled plants will end up impacting public health in poor, minority, or otherwise vulnerable communities.

It's not clear how much of a worry hotspots should be. For instance, a 2018 study found that RECLAIM reduced risks in all segments of the population, with greater reductions for blacks than whites but smaller reductions for Latinos than whites.[3] A previous

---

[2] For a detailed description of the RECLAIM program, *see generally* Daniel P. Selmi, *Transforming Economic Incentives from Theory to Reality: The Marketable Permit Program of the South Coast Air Quality Management District*, 24 ENVTL. L. REP. 10,695 (1994).

[3] See Cobett Grainger and Thanicha Ruangmas, *Who Wins from Emissions Trading? Evidence from California*, 71 ENV. & RES. ECON. 703 (2018).

study found no evidence that changes in emissions were related to demographic characteristics.[4] Still, the possibility that hotspots will harm poor communities or communities of color cannot be ruled out. It is also possible that stringent conventional regulations could secure greater health gains for those communities.

Fourth, even in a well-designed market, the price of allowances may be volatile and unpredictable. For instance, as economic activity declines during a downturn, permit prices will also go down: less electricity is used and factories are producing less, so carbon emissions decline, freeing up more allowances for sale. This unpredictability may make it difficult for businesses to plan.

B. Emissions Trading Versus Taxes

Emissions taxes avoid this unpredictability issue: the government simply sets a price per ton of emissions and sticks with it. For this reason, emissions taxes have some benefits from a business point of view. They also avoid the hotspot problem since every source has an equal incentive to reduce emissions in order to limit its tax payments. But emissions taxes have two major negatives. The first is political: in many places, the idea of a new tax is simply unacceptable to voters. This negative reaction may take carbon taxes off the table, politically speaking. The second is that, although the tax eliminates uncertainty about prices, it creates uncertainty about the extent of emissions reductions. Firms will have to decide when to reduce carbon and when to simply pay the tax. When reducing carbon is cheaper than paying the tax, firms will reduce emissions rather than paying the cost. The problem is that we cannot be sure in advance of the extent to which this will be true, which will depend in part on how ingenious companies are in finding methods of carbon reduction. So the actual amount of carbon reduction may be larger or smaller than predicted.

A carbon tax could be a major source of revenue, raising the question of what to do with the money. One possibility is a consumer rebate. The rebate has to be independent of the amount of energy used by a consumer, because otherwise consumers would have an incentive to increase their energy use rather than becoming more efficient. The rebate could simply be per capita, or it could vary depending on income or some other factor. Alternatively, the funds could be earmarked for special purposes such as financing research on clean energy or projects to reduce the impact of climate change, such as seawalls to guard against sea level rise. Finally, the alternative favored by most economists is to use the revenue to

---

[4] See Meredith Fowlie, Stephen P. Holland, and Erin T. Mansur, *What Do Emissions Markets Deliver and to Whom? Evidence from Southern California's NOx Trading Program*, 102 AMER. ECON. REV. 965 (2012).

reduce other taxes. In particular, many economists disfavor the corporate income tax and would be happy to see a carbon tax replace it as a source of government revenue.

Economists tend to be skeptical of traditional forms of regulation, friendlier toward carbon trading, and most enthusiastic about a carbon tax. Their basic reason for favoring a carbon tax over trading relates to uncertainty. If we had perfect information, a carbon tax and a trading market would be equivalent—there is no difference in outcomes between a trading system and a carbon tax where the tax equals the price that permits would command in the trading system. But in the real world, we are not certain about either the right tax (which would equal the social cost of carbon) or the right level for the cap.

In turn, the two systems create different forms of uncertainty about outcomes. With a carbon tax, the cost of emission reductions is predictable, since businesses will reduce emissions up to the point that further reductions would cost more than simply paying the tax. But the quantity of reductions is uncertain because we do not have perfect information about the costs of future emission reductions for each firm. So a carbon tax creates certainty about costs but not about emission quantities. On the other hand, a carbon market provides certainty about the quantity of emissions, because they will not exceed the cap. But a carbon market creates uncertainty about the cost of emissions reduction because we do not know in advance what the price of permits will be. So the carbon market creates certainty about emission quantities but not about costs.

The question, then, is which form of uncertainty is most serious. Many economists argue that uncertainty about costs is a greater problem—that is, it is better to get a carbon tax wrong than to get the cap wrong in a carbon market. The reason is that an error in the cap will result in having too much or too little emissions until we correct the mistake. But because harm from climate change is due to cumulative emissions, the impact of a short-term mistake on atmospheric concentrations (and hence the amount of harm) will be relatively small. In percentage terms, even a large percentage mistake in the amount of emissions over some period of time will cause a much smaller percentage difference in the amount of harm. But a mistake in the carbon tax translates immediately into compliance costs, so a percentage mistake in the tax will cause a corresponding percentage mistake in emissions control costs. On this basis, it seems worse to make a mistake about price than about quantity in terms of carbon emissions. For that reason, a carbon tax may be better because it fixes the price and only leaves the amount of emissions uncertain.

To take a concrete example, suppose that the United States adopted a carbon tax, but we discovered after five years that the tax was 50 percent too high. That is, we would hypothetically discover that the tax was that much higher than the social cost of carbon. This would make a major difference in the cost of emissions reductions during that time period. Now suppose, on the other hand, that we had a trading scheme and set the cap 50 percent too high for five years. The five-year "excess" emissions will have a very small impact on total atmospheric concentrations, given that concentrations represent the cumulative results of at least 150 years of global emissions. So the amount of harm done by the mistake will be relatively small. Hence, if we have to make a mistake, the consequences will be smaller if we make a mistake about the quantity of emissions rather than the price. This is a somewhat subtle argument, and there are a number of assumptions built into the argument (for instance, that short-term quantity mistakes do not change the long-term emissions trajectory). At least in theory, however, it does support a preference for a carbon tax over a carbon market.

Although this argument may be persuasive to economists, it has probably had less of an impact in the policy arena. If you had trouble following it yourself, imagine trying to explain it to a busy legislator. As a practical matter, the strongest arguments for a carbon tax are pragmatic: collecting a tax may be simpler than running an emissions trading program, and the tax is more transparent. Taxes may also be easier to accommodate under international trade rules, which allow some taxes to be charged to imports and rebated for exports. The most compelling counter-argument is that taxes may be politically infeasible. Also, it may be easier to build coalitions of jurisdictions by linking trading systems than by harmonizing taxes, and trading systems provide an easy way to reallocate costs by changing the system for handing out permits.

In theory, the differences between conventional regulation, emission markets, and emission taxes are clear-cut. In practice, the boundaries may blur. A conventional regulation can allow purchase and sale of compliance credits, thereby taking on some features of an emissions market. The regulation might also allow firms to pay a fixed civil fine for exceeding their limits, providing a tax-like feature.

Similarly, emissions markets can also take on some tax-like characteristics. Rather than giving allowances to firms, the government can auction them off, collecting revenues like a tax. When auctions are used, the government is then faced with the question of what to do with the proceeds. In order to provide more price stability, the government can put a ceiling on allowances costs or a floor below them. When prices are limited in both directions, this is sometimes called a "price collar." A trading scheme with auctioned allowances and a price collar starts to look very much like a tax. So, the choice is really between a market with some tax features (and perhaps some supplementary emissions regulations), a carbon tax (which might also have some supplementary emissions or market mechanism), or a regulation that may have some market or tax features. This reality tends to blur the comparisons.

In general, regardless of whether traditional regulation, carbon trading, or a carbon tax is used, energy prices will probably go up. The reason is that renewables often remain more expensive than conventional energy sources, especially coal, although the gap is closing. If renewables ever become cheaper than all conventional energy sources, and the electric power grid evolves to accommodate more distributed energy sources, regulation presumably will become unnecessary to reduce emissions since the market will force firms in that direction. There are some signs of this trend in places today.

Another important consideration is that the poor spend a greater percentage of their income on energy costs like heating and gasoline than the affluent. Hence, under any regulatory approach to reduction emissions that generates costs, the poor will suffer a disproportionate cost burden. One advantage of cap and trade, along with carbon taxes, is that they generate funds that can be used to offset this increased burden. Other means have to be used to do so under conventional regulation, such as changes in price structures for electricity, tax credits, or increased funding for other government programs aimed at the poor.

Readers will have to decide for themselves which of these systems seems most appealing. But before making a decision, it is helpful to consider the real world experience with market instruments in the context of climate change. Have these schemes worked smoothly and effectively? Have they achieved their carbon reduction goals?

II. Experience with Market Instruments

In the last section we discussed some of the policy issues involved in choosing between regulations, taxes, and trading systems. In this section, we will take a look at efforts to implement these systems. We start first with carbon markets and then turn to carbon taxes.

A. Carbon Markets

Experience with emissions trading dates back to a U.S. cap-and-trade program for sulfur dioxide from power plants. The program proved very successful, reaching the necessary emissions reductions at an unexpectedly low cost. It created a cap-and-trade system for addressing SO_2 emissions, setting the absolute ceiling (the "cap") on emissions by electric utilities nationwide at 8.9 million tons—an estimated 10 million tons below 1980 levels.[5] Congress left the mechanisms for achieving reductions unspecified, allowing individual firms to determine the most appropriate compliance pathway, for example: energy conservation, the use of cleaner fuels, installation of pollution control technology, or purchase of additional allowances. Congress authorized EPA to distribute allowances annually through a combination of mechanisms, including auctions and free allocation to firms.[6] Allowances can be transferred (bought and sold) within the cap.[7] Therefore, firms that are able to reduce their emissions may sell excess allowances, creating an incentive to develop and implement better emissions-control technologies.

From the outset, this SO_2 trading program provoked considerable scholarly discussion and controversy about the program's costs, cost savings, and its environmental and public health benefits. The early history of the Acid Rain Program produced mixed results. For example, trading between companies was initially limited as a result of public utility rules and flaws in implementation, though intra-company trades were more common.

Nevertheless, the general verdict is that the Acid Rain Program has been successful in reducing emissions at low cost. A 2011 data review suggested that SO_2 allowances were much cheaper than originally expected because industry found less expensive ways to reduce emissions, saving up to one billion dollars per year in compliance costs.[8] Price increases for consumers have presumably also been less than expected. To some extent, the program benefitted from fortuitous changes in fossil fuel prices in favor of lower-sulfur coal and natural gas. However, more recent work introduces two caveats: First, price savings were less than they could have been because some firms picked the more expensive compliance route of installing scrubbers in order to benefit from more favorable treatment by state utility regulators. In addition, railroads raised shipping rates to coal-fired plants and were able to capture some of the cost savings.[9]

Experience with other emissions trading programs have also surfaced some recurrent problems. One is price volatility. Businesses need predictable prices in order to plan. Several trading systems,

---

[5] *See* 42 U.S.C. § 7651b(a)(1).

[6] *See id.*

[7] *See* 42 U.S.C. § 7651b(b).

[8] *See* WILLIAM C. WHITESELL, CLIMATE POLICY FOUNDATIONS 165–166 (2011).

[9] Joseph E. Aldy et al., *Looking Back at 50 Years of the Clean Air Act*, 60 J. ECON. LIT. 179, 188–190 (2022).

most notably California's RECLAIM program (discussed above), have hit huge, unexpected price spikes. There are two ways to handle volatility. The direct way is to set a cap on prices, whereby the authority promises to issue additional allowances if the program hits the cap in order to bring the price down. This can be combined with a price floor if desired, whereby allowances are retired if the prices dip too low. An indirect way to deal with the problem is to allow firms to bank allowances that they do not currently need or to borrow allowances from an allowance bank, so that the market itself can smooth out price fluctuations. As experience has grown with trading schemes, designers have developed more sophistication in anticipating possible snags. Still, as the 2008 Financial Crisis showed, markets are always capable of surprises, however well we think we understand them.

Although emissions trading was initially used for conventional pollutants, carbon markets have become increasingly common. Emissions trading was at the heart of the Waxman-Markey bill, the climate change legislation that passed the U.S. House of Representatives in 2010 only to die in the Senate. Undaunted, a number of state governments have inaugurated emissions trading schemes in the United States. The Northeast Regional Greenhouse Gas Initiative (RGGI), which is currently composed of eleven states,[10] created a multistate trading system for power plant emissions with the goal of achieving a 10 percent reduction by 2019.[11] Each state runs its own emissions trading program, but the states all accept allowances issued by the other states. Almost all the allowances are auctioned. The effect is to create a multi-state market. In 2012, the RGGI states adopted a new 2014 cap of 91 million tons. The RGGI CO_2 cap then declined 2.5 percent each year from 2015 to 2020. Current plans call for the cap to fall another 30 percent from 2020 to 2030. As of June 2022, the allowance price was $14, which may not exert strong pressure to reduce emissions. However, the proceeds are used to fund energy efficiency and renewables, which do decrease overall emissions. This is not an insignificant benefit, since total RGGI revenues as of mid-2022 were $4 billion.

Allowance prices are also significant for another reason, since they may allow us to determine whether the system is producing a carbon price that approximates the social cost of carbon. In the case of RGGI, the carbon price has been too low to fully internalize the

---

[10] Regional Greenhouse Gas Initiative, An initiative of the Northeast and Mid-Atlantic States of the U.S., http://www.rggi.org/. The current members are Connecticut, Delaware, Maine, New Hampshire, New Jersey, New York, Rhode Island and Vermont.

[11] *See* http://www.rggi.org.

harm done by carbon emissions. Nevertheless, emissions from RGGI states have decreased markedly.

Another major U.S. trading program was launched in California. California's cap-and-trade program, adopted under a law known as AB 32, sets a declining, statewide cap on greenhouse gas emission and covers about 85 percent of the state's emissions. The cap was set in 2013 at about 2 percent below the emissions forecast for 2012, declining annually until 2020, when the cap was about 20 percent below 2015 levels. The program initially covered only electricity generators and large industrial facilities but in 2015 distributors of transportation, natural gas, and other fuels were added to the system. The program was also expanded to link with Quebec's emissions trading system. At this writing, prices have reached $28 per ton, having risen quickly recently (perhaps due to the disruption of energy markets caused by the war in Ukraine).

By 2017, several proposals to revise the trading system were before the legislature, with sponsors hoping to get two-thirds votes in order to eliminate potential state constitutional challenges. Environmental justice advocates had hoped to move away from cap-and-trade or to dedicate auction revenues to disadvantaged communities, while many conventional environmental groups favor continued use of cap-and-trade. Industry supported cap-and-trade for fear that regulatory alternatives will be more expensive.

In response to these proposals and the shifting legal landscape, California made significant changes in its emissions trading scheme. Two changes are of particular note. In SB 32, the legislature adopted a new target: reducing emissions 30 percent below 1990 levels by 2030. Another new law, AB 197, required regulators to prioritize:

> (a) Emission reduction rules and regulations that result in direct emission reductions at large stationary sources of greenhouse gas emissions sources and direct emission reductions from mobile sources.
>
> (b) Emission reduction rules and regulations that result in direct emission reductions from sources other than those specified in subdivision (a).

A key dispute in the legislative battle involved auction revenues. In 2016, the state invested $1.1 billion from auction revenues in programs such as high-speed rail and sustainable affordable housing. Some legislators wanted to continue this spending agenda. Others would instead have liked a cap-and-dividend system, returning auction proceeds to taxpayers, or would have liked to treat the proceeds as general revenue available to fund all state programs, not just carbon reduction efforts.

On July 17, 2017, the California legislature passed AB 398, which was quickly signed by Governor Jerry Brown. By a two-thirds vote, including a handful of Republicans, the legislature extended the states emissions trading system through 2030, by which time the state must reduce emissions forty percent below 1990 levels. The new law prioritizes expenditure of auction proceeds for reducing non-carbon air pollution as desired by environmental justice advocates, low-carbon transportation options, climate adaptation, and several other purposes.

Other states have also adopted trading programs. Washington State initially took a different path to California by adopting a hybrid of conventional regulation and emissions trading. Legislation in 2021 revamped the system. The new system is modeled on California's, and the state is considering linking its market with California so allowances could be traded freely between states.

The US greenhouse gas market efforts were predated by European efforts. The EU began operating the world's first mandatory carbon trading scheme in January 2005.[12] For internal political reasons, the EU distributed emissions ceilings amongst its member nations, allowing emissions in some countries to grow while others faced sharp reductions. EU members then established their own trading programs, using a variety of schemes to allocate permits to their industries. The program got off to a rocky start. Allowance prices fluctuated from as little as one euro to as much as thirty. The EU made major revisions to the system. Under the new system, the system of national caps has been replaced with a single cap, and auctioning is now the default method for allocating allowances. Before, they were typically just given away to firms.

A further revision to the EU trading scheme took place recently. The new target is a 55 percent cut in emissions from 1990 levels by 2030 and zero-net emissions by 2050. In the summer of 2021, before fear of war in Ukraine began driving prices up, the allowance price was about €60.

Systematic data about national efforts and their effectiveness is not easy to come by. A study by the Australian government of the electricity sector found a range across countries, with Germany and the United Kingdom well in the forefront and other countries showing only minor reductions to estimated emissions compared with business as usual.[13]

---

[12] European Commission, Emissions Trading System (EU ETS), https://climate.ec.europa.eu/eu-action/eu-emissions-trading-system-eu-ets_en [hereinafter European Commission].

[13] Australian Government Productivity Commission, *Carbon Emission Policies in Key Economies* 75 (Table 4.1) (2011), http://ssrn.com/abstract=2006078.

Although developed countries have been most willing to incur costs in mitigating climate change, there has also been some movement among developing countries. At one point, China was considered to be among the countries most resistant to emissions reduction. But China's twelfth five-year plan, which covered 2011–2015, included a 17 percent reduction in energy intensity and a 10 percent increase in the share of energy produced by renewables and nuclear.[14] The plan also established a longer-term goal of establishing a carbon trading market. Implementation of the plan included pilot programs consisting of emissions trading systems in seven Chinese cities. Trading was launched in Shenzhen in June 2013, followed quickly by trading systems in Shanghai, Beijing, Guangdong, and Tianjin.

The requirements covered all enterprises within a region, including emissions from imported electricity. The trading schemes included only large sources in some cities but smaller ones in others such as Shenzhen. There were other differences, too, in the extent to which offsets can be used, in the schemes for allocating allowances, and in techniques to limit price volatility. Various methods were also used to ensure compliance, ranging from monetary penalties to including non-compliance in credit reports to reducing future allowance allocations based on shortfalls. Carbon prices varied, but in Shenzhen reached about $10 per ton. As part of its increasingly ambitious efforts to address climate change, China initially planned to announce a national cap-and-trade market in 2017 but the launch was delayed for almost five years.

In 2021, China launched a national trading system covering four gigatons of carbon annually. China's current climate targets call for peaking CO_2 emissions before 2030; lowering CO_2 emissions per unit of GDP more than 65 percent below 2005 levels by 2030; and achieving carbon neutrality by 2060. The cap "floats" in the sense that it is determined by the carbon intensity of operations (output per ton of carbon burned) rather than the total amount of carbon emitted. Thus, it functions to lower carbon intensity, but the effect on total emissions depends on how quickly GDP grows. Allowances are allocated on the basis of a percentage of 2018 emissions. The theory is that carbon intensity will drop quickly enough to cut total carbon even as total production grows.

The prices produced by existing emissions markets around the world have not been high, perhaps a reflection of their limited degree of ambition (and in some cases, over-allocation of allowances). Nevertheless, they do seem to have had some success in limiting

---

[14] JOANNA LEWIS, ENERGY AND CLIMATE GOALS OF CHINA'S 12TH FIVE-YEAR PLAN 1 (Pew 2011).

emissions. So far, none of them has run into serious operational problems. Thus, they suggest that emissions trading can be a workable approach to emissions reduction, but they also indicate that emissions cuts will be limited without aggressive reductions in the level of caps.

B. Carbon Taxes

Although carbon taxes have not been implemented as broadly as carbon trading, they seem to have worked smoothly in the cases where they have been adopted. In 2008, the Canadian province of British Columbia adopted a climate tax, which is currently $50 (Canadian) per ton. The tax is collected from sellers of fuel, except for natural gas, where it is collected from the user. For instance, the tax on gasoline is about 40¢ (Canadian) per gallon. In order to deal with equity concerns, some of the revenue is used to fund a low-income tax credit. In recent years, carbon emissions have leveled off, and the province seems to be looking to other policies to cut emissions in the future.

Canada also has a national tax, which is a fallback measure applying in provinces that fail to adopt sufficiently rigorous climate policies. The national tax in 2022 was also $50. The tax is designed to be revenue neutral. Canada plans to increase the tax by $15 per year through 2030.

Thus, experience suggests that carbon taxes would indeed be a workable approach to limiting carbon emissions, as their supporters argue. The biggest barrier has been—and seems likely to continue being—political. In most places, politicians seem to be reluctant to support the creation of new taxes, even if the taxes are supposed to be offset by cuts in other taxes. Also, in some jurisdictions, new taxes may face a supermajority requirement, whereas trading schemes only need to clear the normal legislative hurdle.

An additional wrinkle involves protection against competition by firms outside the jurisdiction who may benefit from lower prices than local firms if they are exempt from the carbon tax. To be effective, a carbon tax needs to be imposed on imports, as well as local production, based on their carbon content. Otherwise, reductions in carbon emissions could simply be offset by an increase in cheaper, higher-carbon imports. In the international context, such a tax would be subject to international trade rules, though many experts anticipate that such taxes would be legal. At the state level, the dormant commerce clause is a problem, since such a tax would directly burden interstate commerce. Those issues relating to state regulation are discussed in Chapter 7.

It remains to be seen whether carbon taxes will eventually surmount these barriers. In the meantime, emissions trading seems to be a more popular mechanism for emissions reductions.

III. Credits and Offsets

As a practical matter, neither a carbon tax nor a carbon market is likely to include all sources of carbon. In theory, a trading system could require permits from every source of carbon on the planet, including home fireplaces, forestry companies due to emissions caused by deforestation, and farmers due to methane from animals. Similarly, in theory, each of those emissions might be monitored and taxed directly. In reality, it seems impractical to monitor and tax (or require a permit for) every source of emissions even within a single jurisdiction, and so far, market instruments also have geographic limitations. In addition, besides carbon sources, there are also carbon sinks. For instance, planting trees takes carbon out of the atmosphere.

Although it may be impractical to directly cover every carbon cap and sink, it remains possible, even likely, that carbon reduction can sometimes be obtained more cheaply by looking outside of the sources that are taxed or covered by the trading scheme. "Offsets" allow covered sources to benefit from non-covered emissions reductions or carbon sinks. Those non-covered reductions or sinks can sell offsets or credits to covered sources. The covered sources can then use them in lieu of carbon allowances to satisfy their obligations in a trading system, or they might provide a credit against a carbon tax.

Despite the appeal of offsets as a method of reducing the cost of emission reductions, environmentalists tend to be skeptical of offset programs, and with some justification. One problem is that the offsets may involve activities or locations that are hard to monitor. How can we be sure that newly planted trees will not be logged as soon as the credits are sold, or that a power plant in another country has reduced emissions as much as it claims? Moreover, even if those trees were planted or the emissions were reduced, it's possible that this would have happened anyway, without the incentive of the offset sale. A covered source should not be able to offset its own emissions with carbon reductions that it did nothing to cut. This is called the problem of additionality (making sure that any reductions are "additional" to what would have happened anyway).

In order to counter these concerns, offset programs need to ensure proper monitoring and obtain adequate assurances that reductions are permanent. They also need some method to gauge additionality, which is a difficult task because it requires constructing a counterfactual (what *would have* happened in the absence of the offset program). On the assumption that errors will sometimes be made, offset programs often limit the extent to which offsets can be used by covered sources, or they discount offsets, for instance, by giving covered sources credit for only a percent of the purported emissions reductions from the offset.

The best-known offset program is the Clean Development Mechanism (CDM) under the Kyoto Protocol. One goal of the CDM was to provide a source of funding for developing countries, whose emissions were not capped under the agreement, to engage in voluntary emissions reductions and to facilitate sustainable development. At the same time, emitters in developed countries would gain the opportunity to meet their emission reduction obligations more cheaply, rather than having to invest in more expensive reduction methods at home. Projects in those countries could be submitted for approval, accompanied by certification from third-party verifiers showing that emission reductions will be real and would not have taken place otherwise. Once projects were registered, carbon credits were issued which could be purchased and used by emitters in developing countries to offset their own emission reduction obligations. In addition, a levy was applied to each project and the money that was generated was funneled to the Adaptation Fund, which financed adaptation projects in developing countries that are particularly vulnerable to climate change.

The CDM has been controversial for a number of reasons. In the early days of the CDM, for example, a number of concerns centered around the unequal geographic distribution of projects, with the majority being cited in rapidly developing economies such as China and India, and with very few being located in the Least Developing Countries, where there is an urgent need for investment in sustainable development. The biggest issue relating to many CDM projects was whether they are actually creating emission reductions or giving credit for "anyway" emission levels (either because the baseline was wrong, so cuts were illusory, or because cuts were genuine but would have occurred anyway). Reforms were undertaken to eliminate or at least reduce these problems. Chapter 3 contains a more detailed discussion of CDM and other "flexibility mechanisms" under the Kyoto Agreement and, more recently, the Paris Agreement.

The California cap-and-trade scheme provides another example of an offset program. California has explored the possibility of providing carbon credits ("offsets") to local emitters based on forest preservation in other countries.[15] Under California's AB 32, offsets

---

[15] It should also be noted that the failed effort in 2009–2010 to pass federal climate change legislation featured a similar offset program. *See* Maron Greenleaf,

can be used to satisfy a small part of an in-state source's obligation to cut its carbon emissions. Half of offsets must supply direct environmental benefits to the state. California has detailed guidelines for when forest preservation in the U.S. can be used for offset purposes.[16] The state has investigated the possibility of expanding this program to include forests in Latin America. According to supporters of this effort, the "MOU between California, Acre, and Chiapas represents a historic opportunity to strengthen jurisdictional REDD+ programs, securing and deepening the substantial progress that has already been made in lowering carbon dioxide emissions to the atmosphere associated with tropical deforestation."[17]

The California effort also has outspoken critics. They argue that "expecting carbon markets to do the job of protecting forests can, and does, lead to very problematic outcomes," including destruction of subsistence farming, evictions of underprivileged groups from their lands, and increased social conflict.[18] California's latest legislation preserves offsets but tightens limitations on their use, which seems to have left any cross-border discussions in the lurch.

Use of carbon trading systems to provide incentives for forest preservation has been a lively subject of international interest (and dispute). The negotiations took place under the rubric of REDD+—an acronym that stands for *R*educing *E*missions from *D*eforestation and forest *D*egradation.[19] Negotiators have pursued this goal through a series of international climate conferences. The Copenhagen Accord called for "the immediate establishment of a mechanism including REDD-plus, to enable the mobilization of financial resources from developed countries."[20] The Cancun Agreement went further, requesting a "robust and transparent"

Using Carbon Rights to Curb Deforestation and Empower Forest Communities, 18 NYU ENV. L. REV. 507–511, 523–535 (2011).

[16] California Air Resources Board, Compliance Offset Protocol U.S. Forest Offset Projects, http://www.arb.ca.gov/cc/capandtrade/protocols/usforest/usforest projects_2014.htm.

[17] REDD Offset Working Group, California, Acre and Chiapas: Partnering to Reduce Emissions from Tropical Deforestation: Recommendations to Conserve Tropical Rainforests, Protect Local Communities and Reduce State-Wide Greenhouse Gas Emissions (2013), https://www.foei.org/wp-content/uploads/2018/01/REDD_The-carbon-market-and-the-California-Acre-Chiapas-cooperation.pdf.

[18] Jeff Conant, *California REDD: A false solution*, FRIENDS OF THE EARTH (May 20, 2013), http://www.foe.org/news/archives/2013-05-california-redd-a-false-solution.

[19] *See* David Takacs, *Forest Carbon (REDD+), Repairing International Trust, and Reciprocal Contractual Sovereignty*, 37 VT. L. R. 653, 653 (2013) (explaining the acronym.)

[20] CoP to the UNFCC, Dec. 2/CP.15, Copenhagen Accord, 15th Sess., Dec. 7–19, UN Doc. FCCC/CP/2009/11/Add.1 para. 8 (December 18, 2009). Paragraph 6 defines REDD+ as "reducing emission from deforestation and forest degradation and the need to enhance removal of greenhouse gases by forests."

monitoring system as well as safeguards for sustainable development, poverty reduction, and indigenous peoples. In 2015, finishing touches were put on a rulebook.[21] Implementation of REDD+ has suffered from anemic funding.

Article 5 of the Paris Agreement contains a long, convoluted sentence relating to REDD+. It encourages parties to "support, including through results-based payments," the deforestation decisions and guidance already issued in the UNFCCC process, while also "reaffirming the importance of incentivizing, as appropriate, non-carbon benefits associated with such approaches." The goal seems to have been to endorse implementation of REDD+, but without actually naming it in order to avoid offending some countries.

It remains to be seen what role REDD+ will play in the next rounds of climate negotiations. But the idea seems to have a great deal of momentum. Developed nations had pledged $100 billion by 2020 to support mitigation and activities in developing countries, much of it likely involving REDD+.[22] Deals might be done on a project-by-project basis with local owners, or might involve block transactions with nations or their subdivisions.[23] Five billion dollars in funding has already been pledged specifically to REDD+,[24] and the World Bank has developed a program to help countries prepare to participate in REDD+.[25] In short, the effort seems to be developing some momentum.

While some praise this as an historical opportunity to link climate stabilization and finance for forest protections, REDD+ has been subject to strong criticism, partly by critics who view it as a dodge to enable developed countries to avoid controlling their emissions, and partly by those who feel that it will disadvantage local peoples, who may be forced from their land or otherwise impacted. There are also concerns about the verifiability of carbon reductions. Others argue that REDD+ is a new form of colonialism in which foreigners take control of local lands.[26] (On the other hand, by giving forests financial value, REDD+ might be thought to empower those

---

[21] COP 21, Decision 16/CP.21Alternative policy approaches, such as joint mitigation and adaptation approaches for the integral and sustainable management of forests (Dec. 10, 2015).

[22] Takacs, supra note 19, at 655.

[23] *Id.* at 659.

[24] *Id.* at 664.

[25] Lloyd C. Irland, *"The Big Trees Were Kings": Challenges for Global Response to Climate Change and Tropical Forest Loss*, 20 J. ENV. L. 387, 395 (2010).

[26] *Id.* at 682; Irland, *supra* note 25, at 390. A related argument is that REDD+ would invade national sovereignty. *Id.* at 393. *See also* Keith H. Hirokawa, *REDD+ as the Stranger King*, LPE Project (Dec. 20, 2021), https://lpeproject.org/blog/redd-as-the-stranger-king/.

currently in possession.)[27] But even some critics are beginning to see merit in the idea.[28] There have been efforts to add protections for local populations to REDD+, although at the cost of further complicating transactions and compliance monitoring.

Apart from the considerable problems of verification and monitoring, perhaps the biggest stumbling point with REDD+ is the lack of clarity over who is actually entitled, through possession, ownership, or sovereignty, to make deals for carbon credits and to receive the benefits. In developed countries, formal ownership of land is likely to be clearly specified, squatters have limited legitimacy, and the prerogatives of lawmakers are well delineated. But in developing countries, none of these may hold true, with land occupied by individuals with no title but with other claims of legitimate possession. Land ownership is often contested, and the formal and actual power of lawgivers is much in dispute. By favoring any one of these parties, REDD+ risks intruding into a contested internal matter. Doing so creates practical problems while at the same time invading national sovereignty to the extent that national governments are not allowed to simply settle these issues themselves. Nuanced solutions would be ideal, except for the fact that they further complicate the design and operation of the REDD+ system. It is also important to remember that the alternative to REDD+ is not necessarily preservation of forests and their human inhabitants in their current state; it is damage to the forests from climate change and continued pressures for deforestation by developers.

Despite these difficulties, something like REDD+ seems inevitable for a variety of reasons. From the point of view of the globe at large, it provides an attractive way to reduce carbon emissions at relatively low cost.[29] Developed countries are likely to demand some effort by developing countries to control emissions relating to land use, as a quid pro quo for decarbonizing their energy systems. And REDD+ provides financial benefits that could flow to local residents, landowners (not necessarily the same group), and national governments. The upshot may well be a messy compromise that is just barely satisfactory to enough parties to get the deal done. Or maybe not—the chapter on international negotiations has more to say about the barriers facing negotiators.

---

27 Takacs, *supra* note 19, at 695.

28 *Id.* at 662.

29 *See* Irland, *supra* note 25, at 392 (projected offset cost is lower than mitigation costs in industrial countries).

IV. The Problem of Leakage

Offsets are a response to the incomplete geographic coverage of emissions reduction schemes, particularly in the developing world. That incomplete coverage has other implications as well, one of which is the problem of carbon leakage discussed in this section.

Reducing the use of carbon fuels can have market consequences that can amplify emissions elsewhere. The simplest example involves energy efficiency measures, where the problem is called the rebound effect. The rebound effect can take two forms, as illustrated by improvement in the fuel efficiency of vehicles. First, because it takes less fuel to drive an additional distance, improved fuel efficiency can result in an increase in driving, in turn causing the use of additional fuel. The carbon from the increased driving offsets some of the gains from improved efficiency. Second, because drivers with efficient cars need less fuel, fuel demand is reduced, which causes a price reduction. But the price reduction may cause other drivers to increase their fuel use. Because there is a global oil market, the lower prices can potentially increase car use even in other parts of the world. In the automobile setting, the rebound effect does not seem to be very large, because gasoline demand is relatively inelastic, meaning that a price decrease does not cause much in the way of additional consumption. But the rebound effect may be greater in other settings.

The rebound effect involves increases in energy use that are paradoxically caused by improved energy efficiency. Because of the rebound effect, it is even theoretically possible that increased energy efficiency could actually lead to greater total consumption of energy. But the empirical evidence suggests that the effect is limited. A recent review of the literature concludes that "[a]fter four decades of studies on this phenomenon, there is a strong consensus among scholars as to the existence of rebounds, but there is still no consensus on its magnitude or its real impact on sustainability efforts."[30] The same article concluded that taxes on energy production and pricing carbon can be effective policies, along with efforts to move consumption away from goods to services.

Imposing a price on carbon can also lead to increased emissions elsewhere. One mechanism involves the same kinds of impacts on global fuel markets that can cause rebound. Another mechanism is that higher energy costs might cause industry to move elsewhere, or might cause domestic industry to lose market share to industries in countries that do not price carbon. In 2017, the IPCC cited a leakage

---

[30] Jaume Freire Gonzales and Mun S. Ho, *Policy Strategies to Tackle Rebound Effects: A Comparative Analysis, Ecological Economics* 193 (2022) 107332, https://doi.org/10.1016/j.ecolecon.2021.107332.

estimate of 12 percent (in a range from 7–19 percent) for leakage due to restrictions on the energy system, raised by about one-third if leakage from industrial processes is added.[31] A 2020 EU report found that:

> Ex-ante predictions by simulation models indicate that direct leakage is indeed likely. Its size depends on the difference between the EU's carbon prices and those of its trading partners. On average, studies indicate that about 15% of domestic emission savings are offset by additional foreign emissions. However, the range of estimates is very large. In most studies, indirect carbon leakage that operates through global markets for fossil fuels, however, is quantitatively more important than direct carbon leakage operating through international markets for goods and services.[32]

Thus, leakage seems to be significant but limited, although there is a great deal of uncertainty about the issue.

Leakage is countered to the extent that mitigation efforts lead to technological advances that are adopted elsewhere. The interaction between technology change and impacts on energy prices (the price effect) is somewhat understudied. The limited existing research on technology effects and leakage indicates that the key factors are the elasticity of demand, scarcity of fossil fuels, and substitutability between clean and carbon-based energy.

Regulators can also reduce leakage by excluding industries that compete with foreign firms from the trading system or by granting rebates of carbon taxes or auction fees, which can be designed to provide an incentive for improved performance. When the regulating jurisdiction is also a major importer, border measures can discourage leakage, with particularly strong effect if producers maintain unified production lines. Leakage can also be reduced if other countries anticipate that they will become subject to binding limitations in the future or can sell offset credits to countries that are controlling emissions. Expanding the economic size of the coalition engaged in abatement also decreases leakage. It should also be possible to reduce leakage by taking steps to reduce the supply of fossil fuels, such as restrictions on coal mining or offshore drilling.

---

[31] Gabriel Blanco et al., *Drivers, Trends and Mitigation*, in O. EDENHOFER ET AL., CLIMATE CHANGE 2014: MITIGATION OF CLIMATE CHANGE, CONTRIBUTION OF WORKING GROUP III TO THE FIFTH ASSESSMENT REPORT OF THE INTERGOVERNMENTAL PANEL ON CLIMATE CHANGE 386 (2014).

[32] EU Directorate-General for External Policies Policy Department, *Briefing Economic Assessment of Carbon Leakage and Carbon Border Adjustment* 2 (2020).

Leakage problems will be directly relevant when we discuss state and national carbon restrictions in the next chapters. The effectiveness of those restrictions is limited by leakage concerns—at least, unless they turn out to provide stepping-stones toward a broader global regime. Nevertheless, even these sub-global regimes may have direct climate benefits if leakage is not too severe.

What conclusions can be drawn about pricing carbon? First, economic theory does suggest the carbon-pricing techniques we discussed in this chapter are more cost-effective than conventional regulations. Their actual effectiveness, however, may turn on the details of how they are designed. Second, whether we use carbon pricing or conventional regulation, a key issue is whether we have set a high enough explicit or implicit price of carbon. If a tax or cap is too low, or conventional regulations are too lax, emitters will still be able to impose externalities on the world today and on future generations. Third, current experience shows that trading schemes can be workable, although getting the design right is not always easy. Taxes have been used to a lesser extent but have seemed to function smoothly where they have been adopted. Fourth, no matter what system we use, effective efforts to reduce emissions will ultimately require broad international participation.

When we turn to the U.S. situation, the prospects for national regulatory legislation seem dim at least in the near future, and for that reason, a national carbon tax or emissions trading scheme seems unlikely. As we have seen in this chapter, some states have adopted emissions trading, and it is possible that those schemes will expand to encompass most states. But much of the "action" in the United States involves conventional regulations, despite the preference of economists for market solutions. Recently, the use of large financial subsidies has found favor with Congress, though economists consider subsidies too poorly focused to be efficient. Chapter 6 will consider regulatory efforts at the national level, while in Chapter 7 we will consider how state governments have approached emissions reductions. First, however, we will take a close look at the complexities involved in carbon mitigation in the next chapter.

Further Readings

Bruce Ackerman and Richard B. Stewart, *Reforming Environmental Law*, 27 STAN. L. REV. 1333 (1985).

Reuven S. Avi-Yona and David M. Uhlman, *Combating Global Climate Change: Why A Carbon Tax is a Better Response to Global Warming Than Cap and Trade*, 28 STAN. ENV. L. REV. 3 (2009).

DAVID M. DRIESEN, THE ECONOMIC DYNAMICS OF ENVIRONMENTAL LAW (2002).

Robert W. Hahn and Gordon L. Hester, *Marketable Permits: Lessons for Theory and Practice*, 16 ECOLOGY L.Q. 361 (1989).

International Carbon Action Partnership, Emissions Trading Worldwide: Status Report 2022 (2022), https://icapcarbonaction.com/en/publications/emissions-trading-worldwide-2022-icap-status-report.

Howard Latin, *Ideal Versus Real Regulatory Efficiency: Implementation of Uniform Standards and Fine-Tuning Regulatory Reforms*, 37 STAN. L. REV. 1267 (1985).

Yoram Margalioth, *Tax Policy Analysis of Climate Change*, 64 TAX L. REV. 63 (2010).

Gilbert E. Metcalf and David Weisbach, *The Design of a Carbon Tax*, 33 HARV. ENV. L. REV. 499 (2009).

Bryan Swift, *U.S. Trading: Myths, Realities, and Opportunities*, 20 NAT. RES. & ENV. 3 (2005).

ZHONG XIANG ZHANG, CROSSING THE RIVER BY FEELING THE STONES: THE CASE OF CARBON TRADING IN CHINA (Fondazione Eni Enrico Mattei 2015).

Chapter 5

TOOLS FOR REDUCING EMISSIONS: ENERGY REGULATION

This chapter will examine the relationship between energy use and climate change. We will focus primarily on the way in which electricity and transportations systems are evolving in the United States in response to both market forces and environmental concerns, including but not limited to climate change. Because electricity and transportation systems involve a complex interplay between federal, state, and local law, we will focus on understanding the key characteristics of each of these systems, including the key ways in which different levels of law and policymaking intersect. Sharp divides between the two political parties make energy law an area of law highly susceptible to flux in the coming years.

Energy law is a complex and expansive area of federal and state law. In order to contextualize the discussion of the intersection between energy and climate change, this Chapter will begin by reviewing the primary sources of energy in the United States and discussing how the energy portfolio of the United States has and is likely to evolve in response to climate concerns. We will then briefly review how the United State electricity system operates and how it has evolved over the past forty years in response to market and environmental concerns. After outlining the electricity system, we will discuss different ways in which the energy system can operate to facilitate a shift towards a low carbon economy. Several state-based case studies will be introduced as tools for thinking about different modes of energy shifting. Finally, before moving on to examine the transportation system, we will review how federal land management and offshore oil policies shape access to energy resources, with resulting impacts on climate change.

After exploring sources of energy and the evolving electricity system, the Chapter will introduce readers to the ways in which decision making in the transportation sector influences efforts to mitigate climate change. Here, we will focus on the roles that federal automobile regulation, shifts in the auto fleet, and sub-national city planning, and public transportation policies play in shaping climate-limiting efforts.

The energy sector encompasses a major portion of the American economy, and it is the subject of a correspondingly complex web of state and federal laws. At many law schools, Energy Law is the subject of a separate course. We make no pretense of covering these areas in detail. Instead, our goal is to provide an overview of the key feature most directly related to climate change mitigation. This will set the stage for discussions in Chapters 6 and 7 of federal and state regulations of carbon emissions.

I. Introduction to Energy Policy and Energy Sources

One of the most critical challenges facing the United States and the global community is how to ensure access to safe, reliable, and clean sources of energy. Energy use has been fundamental to human development, with fossil fuels having driven the industrial revolution and having been seen as essential to continuing patterns of human development and prosperity—and with access to energy, generally, being seen as critical to social and economic development worldwide. Further, with global annual energy investments valued at approximately $1.9 trillion, the domestic energy sector estimated to be 6.2% of the country's GDP (roughly a $13 trillion value),[1] and with international access to energy stalled but set to expand,[2] the energy industry constitutes one of the pillars of the domestic and international economies. Because our sources of energy are derived from natural resources that must be extracted or otherwise harnessed, transported, and often processed, however, the energy sector is also one of the greatest sources of environmental externalities worldwide. Fossil fuels, in particular, are associated with the creation of externalities at every point in their life cycle—that is, from the point of extraction, through their transportation, to combustion and disposal.

As a result, many early environmental laws and policies focused on minimizing the environmental externalities associated with fossil fuels, particularly coal, oil, and gas. As the links between fossil fuel combustion and climate change became apparent, efforts to limit fossil fuel consumption became one of the primary pillars of climate mitigation strategies worldwide. Continuing reliance on fossil fuels for the brunt of global energy production constitutes one of the most significant challenges to meeting the 2° goal of the Paris Agreement.

---

[1] LINCOLN L. DAVIES, ALEXANDRA B. KLASS, HARI M. OSOFSKY, JOSEPH P. TOMAIN, ELIZABETH J. WILSON, ENERGY LAW AND POLICY 12 (3rd Ed. 2022).

[2] International Energy Association (IEA), *World Energy Outlook* 27, 175 (2021), https://iea.blob.core.windows.net/assets/4ed140c1-c3f3-4fd9-acae-789a4e14a23c/WorldEnergyOutlook2021.pdf (noting the UN Sustainable Development Goal of achieving universal energy access, but finding that as a result of the COVID-19 pandemic "the global number of people without access was broadly stuck at where it was between 2019 and 2021, after improving 9% annually on average between 2015 and 2019. In sub-Saharan Africa the number of people without access increased in 2020 for the first time since 2013."); EIA, *EIA Projects Nearly 50% Increase in World Energy Use by 2050, Led by Growth in* Asia (Sept. 24, 2019), https://www.eia.gov/todayinenergy/detail.php?id=41433.

The International Energy Agency succinctly sums up the centrality of the energy sector to addressing climate change and meeting the 2° goal:

> The energy sector is responsible for almost three-quarters of the emissions that have already pushed global average temperatures 1.1°C higher since the pre-industrial age. . . . The energy sector has to be at the heart of the solution to climate change. At the same time, modern energy is inseparable from the livelihoods and aspirations of a global population that is set to grow by some 2 billion people to 2050, with rising incomes pushing up demand for energy services, and many developing economies navigating what has historically been an energy- and emissions-intensive period of urbanisation and industrialisation. Today's energy system is not capable of meeting these challenges; a low emissions revolution is long overdue.[3]

Before looking briefly at the key energy sources, it is worth noting that energy sources are generally divided into two categories: renewable and nonrenewable sources. Renewable and nonrenewable sources of energy can both be used to produce secondary energy sources such as electricity or hydrogen for transportation fuel. Despite growth in the renewable energy sectors, nonrenewable sources of energy in the form of fossil fuels continue to provide the majority of the world's energy supply. Despite steep declines in recent years, coal continues to provide the largest share of energy for global electricity generation. Moreover, coal began to experience a rebound in demand in 2021 in the wake of the global pandemic.[4] At the domestic level, energy consumption has largely flattened out, with the majority of changes in energy consumption being among the sources being used. In particular, U.S. coal consumption has decreased as coal has increasingly lost market share to the natural gas and renewable generation sectors, both of which have grown rapidly over the past decade, with natural gas use growing more than any other fuel source in terms of the sheer amount of energy consumed, but with renewable energy growing the fastest on a percentage basis.

---

[3] *World Energy Outlook, supra*, at 15.

[4] IEA, *Global Energy Review 2021: Coal* (2021), https://www.iea.org/reports/global-energy-review-2021/coal. The global pandemic coupled with the Russian invasion of Ukraine has put major strains on parts of the modern energy system, leading to sharp price increases and significant rebounds in coal and oil use. As a result, 2021 saw the second-largest annual increase in carbon dioxide emissions in history. *World Energy Outlook, supra* note 2, at 15.

A. Energy Sources: Nonrenewable Sources

Nonrenewable fossil fuel energy sources have been the predominant source of domestic and global energy use for well over a century, and still account for 86 percent of total world energy consumption and roughly 87 percent of domestic energy consumption.[5] The primary sources of nonrenewable energy that are used in the United States include: coal (11%), petroleum (36%), natural gas (32%), and uranium/nuclear electric power (8%).[6]

Historically, coal has played a pivotal role in human development. Coal is available throughout the world, with heavy concentrations in the United States, Australia, and China. Coal, which consists of decomposed and compressed plant and animal matter, is found in seams of sedimentary rock. It can be extracted from underground or surface mines and is ranked based on its heating value and carbon content. The extraction and combustion of coal has long been recognized as environmentally intensive, but coal now also recognized as one of the most climate-intensive sources of energy.

Different fuels emit varying amounts of carbon dioxide in relation to the energy they produce when burned. As a result, the carbon intensity of fuels can be ranked by comparing the amount of CO_2 emitted per unit of energy output (or heat content).[7] Based on this metric, coal is the most CO_2 intensive source of widely used energy, meaning that it is one of the primary contributors to rising levels of CO_2 in the atmosphere.[8] Coal's large footprint is largely due to the widespread use of coal in electricity production. Coal-fired power plants have played a pivotal role in the development of the U.S. electricity system and, despite rapid declines in recent years, continue to produce roughly 22 percent of domestic electricity.

Coal is particularly problematic from a climate perspective because of two intersecting factors: (1) it is CO_2 intensive; (2) it is a relatively cheap form of energy. Its carbon-intensity makes it a key culprit in contributing to climate change, but its low-cost makes it appealing as a source of cheap energy worldwide. The majority of ongoing energy growth is taking place in the developing world, particularly in the rapidly developing economies of China, India,

---

[5] IEA, *Key World Energy Statistics 2021: Supply*, https://www.iea.org/reports/key-world-energy-statistics-2021/supply (this includes nuclear energy).

[6] EIA, *U.S. Energy Facts Explained*, https://www.eia.gov/energyexplained/us-energy-facts/.

[7] The resulting number is known as the fuels Global Warming Potential (GWP). UNFCCC, *Global Warming Potentials*, http://unfccc.int/ghg_data/items/3825.php.

[8] EIA, *How Much Carbon Dioxide is Produced when Different Fuels are Burned?*, https://www.eia.gov/tools/faqs/faq.php?id=73&t=11.

Brazil, South Korea, and South Africa. Rapid growth of coal-fired power energy in China has been of particular concern since China is both the largest net global greenhouse emitter and the world's largest consumer of coal, making China's domestic energy policy choices critical to global efforts to constrain climate change.[9] In many of the areas where new coal-fired power plants are being constructed, improved access to energy will provide economic opportunities that can help reduce poverty rates and help standards of living. If, however, these growing energy demands continue to be met through the proliferation of coal-fired power plants, it will be extremely difficult to constrain global climate change. As a result, linkages between energy and human development sit at the very heart of the climate change debate.

In the United States, coal has rapidly been losing ground to natural gas, another nonrenewable resource. Natural gas primarily consists of methane (CH_4). Natural gas is largely produced from wells devoted to natural gas extraction, through co-production with oil, or from coal beds as coal bed methane.[10] In contrast to coal, natural gas is seen as a cleaner burning alternative that produces fewer environmental externalities. From a climate change perspective, natural gas has less global warming potential than coal but, because methane acts faster but lasts less time in the atmosphere than does carbon dioxide, methane leaks can have greater short-term, direct effects on warming.[11]

Natural gas use has been on the upswing in the United States since 2005 as a result of shale gas extraction and hydraulic fracturing. The development of hydraulic fracturing (fracking) technology, combined with horizontal drilling, has vastly expanded the number of active natural gas wells in the United States. As a result of these new wells, natural gas has flooded the market, driving down the price of natural gas and making natural gas more cost competitive with coal in the energy context. Consequently, as discussed *infra*, natural gas use has increased more than any other fuel source in terms of the quantity of energy consumed, driven in key part by electricity sector, where natural gas has displaced coal as the largest source of electricity generation. Natural gas is now used to produce roughly 38 percent of domestic electricity.[12]

---

[9] Lydia McMullen-Laird et al., *Air Pollution Governance as a Driver of Recent Climate Policies in China*, 9 CARBON & CLIMATE L. REV. 243, 246–47 (2015).

[10] DAVIES ET. AL., *supra* note 1, at 101.

[11] EPA, *Understanding Global Warming Potentials*, https://www.epa.gov/ghgemissions/understanding-global-warming-potentials.

[12] EIA, *What is U.S. Electricity Generation by Source*, https://www.eia.gov/tools/faqs/faq.php?id=427&t=3.

The other source of nonrenewable energy that can be extracted through fracking is crude oil. Oil, which can be recovered from onshore and offshore wells, can be used for heating and electricity, but is largely desirable as a transportation fuel. Crude oil can be refined to produce a number of useful fuels, including gasoline (petrol), diesel oil, fuel oil, and kerosene. Given its many end uses and its relative scarcity, oil is a "globally traded commodity, a cornerstone of world energy and economic markets."[13] From a climate change perspective, oil is most closely associated with the transportation sector and the combustion of petroleum and diesel in automobiles, which will be discussed *infra* in Part III.

The final form of nonrenewable resource that figures large in the climate debate is uranium, or nuclear energy. Nuclear energy uses uranium to produce electricity. Because nuclear energy relies on materials that are used in nuclear weapons and have also been associated with major disasters such as Chernobyl and Fukushima, it gives rise to strong political disagreements.[14] From a climate change perspective, nuclear energy is fairly benign because it produces very few greenhouse gas emissions. Despite initial hope in the 1970s, that nuclear energy could provide cheap, safe, and abundant energy in the United States, as Lincoln et al. describe:

> Significant cost overruns in the construction of planned nuclear plants, overestimates of how much energy the nation would use, increasing regulation of atomic energy production, and the 1979 incident at the Three Mile Island facility in Pennsylvania brought construction of new nuclear power facilities in the United States 'to a screeching halt'".[15] As a result, there has been virtually no growth in nuclear power for over three decades. Globally, however, nuclear energy has expanded. Despite safety considerations, concerns about climate change have led some environmental groups in the United States and abroad to drop their previous opposition and begin to support nuclear energy as a lower-carbon alternative to coal, oil, and gas.[16]

Thus, the future of nuclear energy in the United States is uncertain, but is likely to be driven by market, as opposed to climate forces.

---

[13] DAVIES ET. AL., *supra* note 1, at 98.

[14] Lincoln L. Davies, *Beyond Fukushima: Disasters, Nuclear Energy, and Energy Law*, 2011 BYU L. REV. 1937, 1965 (2011).

[15] DAVIES ET. AL., *supra* note 1, at 106 (quoting Dan C. Perry, *Uranium Law and Leasing*, 55 RMMLF-INST 27-1 (2009)).

[16] Amy Harder, *Environmental Groups Change Tune on Nuclear Power*, THE WALL STREET JOURNAL (June 16, 2016), https://www.wsj.com/articles/environmental-groups-change-tune-on-nuclear-power-1466100644.

B. Energy Sources: Renewables

The primary sources of renewable energy that are used in the United States include: hydropower, solar, wind, geothermal, and biomass. Historically, with the exception of hydropower, renewable sources of energy played a very limited role in the domestic energy sector. More recently, however, our energy system has been shifting towards greater use of renewable sources of energy. This shift is due to several factors, including: environmental externalities, climate change, finite supplies of energy resources, increasing price competitiveness, geopolitical and security concerns.[17] As a result of the confluence of a number of factors, beginning in the 2000s, the development and use of renewable energy began to grow rapidly such that between 2000–2009, the use of non-hydro renewables in the United States more than doubled from 1.9 percent in 2000, to 4.7 percent in 2009.[18] And by 2021, renewables accounted for 12% of domestic energy consumption.[19] Globally, there has been a similar growth in renewable energy, with the IEA reporting that renewable electricity growth is accelerating faster than ever and "[r]enewables are set to account for almost 95% of the increase in global power capacity through 2026", with "[t]he amount of renewable capacity added over the period of 2021 to 2026 is expected to be 50% higher than from 2015 to 2020."[20]

While the various forms of renewable energy have different characteristics, they share some commonalities. By and large, sources of renewable energy tend to have fewer environmental externalities and to have very low greenhouse gas emission profiles. One challenge that different forms of renewable energy share, particularly wind and solar energy, is intermittency.[21] That is, unlike coal or gas that can be kept on hand, readily available to meet varying levels of demand, with wind and solar there is always some level of uncertainty as to when and how much of the resource will be available. This intermittency creates challenges for ensuring that there is a base-load supply of energy to meet varying demands across

---

[17] K.K. DUVIVIER, THE RENEWABLE ENERGY READER 6–13 (2011).

[18] *Id.* at 14.

[19] EIA, *U.S. Energy Consumption by Source and Sector, 2021*, https://www.eia.gov/totalenergy/data/monthly/pdf/flow/total-energy-spaghettichart-2021.pdf.

[20] IEA, Renewable Electricity Growth is Accelerating Faster Than Ever Worldwide, Supporting the Emergence of the New Global Energy Economy (Dec. 21, 2021), https://www.iea.org/news/renewable-electricity-growth-is-accelerating-faster-than-ever-worldwide-supporting-the-emergence-of-the-new-global-energy-economy.

[21] *See, e.g.*, Felix Mormann, *A Tale of Three Markets: Comparing the Renewable Energy Experiences of California, Texas, and Germany*, 35 STAN. ENVTL. L.J. 55, 60 (2016) (noting that: onshore wind and solar photovoltaic (PV) technologies. . .both have recently exhibited the highest growth rates among renewables and, due to their intermittency, present the greatest challenges for successful grid integration).

time. One of the challenges with renewables, thus, is to develop more effective and efficient storage mechanisms, such as batteries, so that any excess energy that is not used at the time it is available can be captured and stored for subsequent use. Another challenge to the widespread, large-scale deployment of renewable energy is the cost effectiveness of different forms of energy. While certain forms of renewable energy, such as wind power, are increasingly cost competitive with fossil fuel sources such as coal and gas, many forms of renewable energy, including solar energy, continue to be more expensive per megawatt hour than traditional fossil fuel sources.[22] The cost competitiveness of different energy sources is often partly attributable to the availability, or absence of availability of subsidies, tax credits, or other governmental incentives. For example, as Outka describes in broad terms:

> Around the world, the International Energy Agency estimates government subsidies for fossil fuels topped four hundred billion dollars in 2010. The effect of such subsidies not only keeps prices artificially low, but also affects energy consumption; the World Bank has reported that eliminating them would decrease energy use by 13% and reduce CO_2 emissions by 16%.[23]

As a result of such subsidies and incentives, efforts to encourage shifts in the domestic and international energy system must consider not only the specific attributes of individual energy sources, but also how existing governmental systems creates opportunities for and barriers to the growth of different forms of energy.

The oldest and most commonly used source of renewable energy is hydropower, which harnesses the power of water through dams and, more recently, through tidal, current, and thermal projects, to generate energy. Concerns about the impact of dams on both natural and human systems, as well as limits to the number of dam-able water systems, has slowed the pace of dam construction and diverted attention towards newer forms of hydropower, such as tidal energy. In contrast to other forms of renewable energy, hydropower has largely remained flat in terms of growth.

In common with hydropower, wind energy has been harnessed for millennia, with early windmills being used as early as 200

---

[22] However, the EIA projects that the costs of certain renewable electricity generation (*e.g.*, wind, solar, hydroelectric) to level out with the cost of other forms of electricity generation (*e.g.*, coal, gas, nuclear, geothermal, biomass) within the next five years. U.S. Energy Information Administration, *Levelized Costs of New Generation Resources in the* Annual Energy Outlook 2022 (March 2022), https://www.eia.gov/outlooks/aeo/pdf/electricity_generation.pdf.

[23] *See, e.g.*, Uma Outka, *Environmental Law and Fossil Fuels: Barriers to Renewable Energy*, 65 VAND. L. REV. 1679, 1696–97 (2012).

B.C.E.[24] In contrast to hydropower, wind energy has experienced significant growth over the past two decades. As DuVivier describes it:

> Wind power development has progressed at breakneck speed in recent years. In installed wind capacity in the United States increased almost fourteen times between 2000 and 2009. U.S. wind capacity increased 50 percent in 2008 alone when 8,500 new MW of wind power generation were constructed nationwide.[25]

As a result of this growth, by 2021, wind energy accounted for 9.2 percent of electricity generation.[26] The growth in wind energy is driven in significant part by its increasing cost competitiveness, aided by the federal Renewable Electricity Production Tax Credit (PTC).[27] The PTC, which was originally enacted in 1992 and has been renewed and expanded in the following years, provides a tax credit for electricity that is generated by certain qualified energy resources. The credit extends for 10 years from the date that the qualifying facility begins operating. Although the level of the available PTC is declining each year, it has, and continues, to serve an important role in bolstering wind energy development.

Wind power, like solar power, is "infinitely renewable" and is associated with minimal greenhouse gas emissions, most of which are associated with the construction of the actual turbine. As a result of its low carbon footprint, ready availability, and cost competitiveness, wind energy is now one of the fastest growing sources of new energy capacity in the United States and worldwide. The majority of wind energy capacity is in the form of terrestrial wind turbines, although there are a growing number of planned or proposed offshore wind energy projects. Offshore winds tend to be more reliable and stronger. Despite the rapid expansion of wind power, further growth is complicated by the intermittent nature of wind energy, challenges with siting wind facilities, and concerns about the impact of wind turbines on wildlife. With respect to the consistency of wind energy, the challenge, of course, is that sometimes the wind blows, and sometimes it does not, so developing more sophisticated energy storage devices is essential to allowing the continued growth of wind energy as an essential part of the U.S. electricity system.

---

[24] DAVIES ET. AL., *supra* note 1, at 117.

[25] DUVIVIER, *supra* note 17, at 73.

[26] *What is U.S. Electricity Generation by Energy Source?*, *supra* note 12.

[27] Energy.gov, *Renewable Electricity Production Tax Credit (PTC)*, https://www.epa.gov/lmop/renewable-electricity-production-tax-credit-information. *See also* Congressional Research Service, *The Renewable Electricity Production Tax Credit: In Brief* (April 29, 2020), https://sgp.fas.org/crs/misc/R43453.pdf.

Siting wind facilities is similarly challenging. First, there are a limited number of prime wind sites. Additionally, many of sites with the greatest capacity to produce high levels of wind power are in relatively remote areas. In order for these sites to be developed, new high voltage electricity transmission lines must be built, which entails substantial construction costs, a complex permitting process, and sometimes, high fees for access to existing transmission lines. In addition, for both terrestrial and offshore facilities, there is often local opposition to the construction of both the turbines and the transmission lines for aesthetic reasons or because of questionable beliefs about their dangers. This is particularly true for offshore locations, such as the Great Lakes, where the proposed sites are located in recreational areas. Equally, the onshore siting process itself is highly complex and involves navigating state laws concerning who owns the wind rights; that is, the question of whether wind rights rest with the owner of the surface estate or whether they are severable in the same way that some mineral resources are. The private property dimensions of wind energy are novel and only a few states have comprehensively addressed wind ownership questions.[28] Finally, even once an optimal site has been identified and property ownership rights have been settled, the permitting process for developing the wind farm is often highly complex and may involve different levels of review by local, state, and federal entities. Because wind energy is still relatively novel, there are significant variations in how different states and localities are responding to proposed projects, with some localities seeking to ban all wind energy development, and with a handful of states, including Wisconsin, Washington, and Ohio, enacting legislation to try to promote wind power development by restricting local control.[29] Still, wind energy has grown quickly and is especially important in a belt running north from Texas through states like Iowa and Kansas.

A final question associated with wind energy development is the extent to which wind projects negatively affect wildlife, particularly in otherwise remote or recreational areas that might be home to abundant and, at times, rare species. Particular concerns include the impact of the development process, including the presence of humans, dust, and noise, as well as the effect of "shadow flicker" from the rotating blades of a turbine, which is thought to "simulate the approach of avian predators."[30] In addition, there are also concerns about birds and bats being killed as a result of flying into turbine blades. As Freemen and Kass explain:

---

[28] DUVIVIER, *supra* note 17, at 86.

[29] *Id.* at 123.

[30] *Id.* at 89.

> [T]he most common impacts are to bird and bat species and include collisions, electrocution, habitat removal, habitat fragmentation, and displacement. It is estimated that approximately 10,000 to 40,000 birds are killed each year by the wind turbines currently operating in the United States.[31]

While the number of wind energy related bird deaths each year may pale in comparison to the 40–50 million bird deaths each year that are attributed to automobiles, the problem is serious and one that has to be evaluated and responded to at each proposed wind site. This is particularly true if there are species in the area that are protected under the federal Endangered Species Act, or other federal laws such as the Migratory Bird Treaty Act or the Bald and Golden Eagle Protection Act.[32] As a result, wind project developers must consult with relevant state and federal agencies in order to understand the relevant wildlife issues and legal constraints that attend to their particular project.

Alongside wind, solar energy is one of the fastest growing sources of renewable energy. While solar energy only accounted for 2.8 percent of U.S. electricity generation in 2021, it is estimated that the percentage share will grow significantly over the next two to three decades, with the EIA estimating that "distributed generation technologies such as solar PV will grow to supply 8% of electricity consumed in households and 6% of electricity consumed in commercial buildings in 2050."[33] Utility scale solar may be an even larger factor.

Solar energy, of course, is energy directly derived from the sun. There are different ways to harness the sun's energy, with photovoltaic (PV), solar thermal, and concentrated solar thermal, being among the three most common techniques. Photovoltaic, or solar PV, are typically made from crystalline silicon or "thin film" semiconductors, and they work by transforming the sun's energy directly into electricity.[34] Solar PV typically takes the form of roof- or ground- mounted panels that can be used individually or as part of a utility scale installation. In contrast, solar thermal operates by transforming the sun's energy directly into heat, or thermal energy, as opposed to electricity. Forms of solar thermal energy include

---

[31] Roger L. Freemen & Ben Kass, *Siting Wind Energy Facilities on Private Land in Colorado: Common Legal Issues*, 39 THE COLORADO LAWYER 43, 46 (May 2010).

[32] *See, e.g.*, Hope Babcock, *How to Choose Between Environmentally Positive Actions When One of Those Actions Can Harm the Other: A Case Study of the Conflict Between the California Condor and Wind Turbines*, 52 ENVTL. L. 1, 3 (2022).

[33] *Id.* at 72.

[34] *See* DAVIES ET. AL., *supra* note 1, at 121.

passive and active solar thermal systems. Passive solar thermal systems focus on orienting buildings to take maximal advantage of available sunlight, while active solar thermal systems use pumps and fans to circulate heat that is captured using some form of collection system, such as a water tank. Finally, concentrated solar thermal (CST) systems employ a large-scale array of devices that "concentrate the sun's heat to generate steam to produce electricity."[35]

Solar PV is one of the newer and fastest growing forms of energy. It is increasingly seen as a desirable form of energy because it is clean, producing very few conventional pollutants or greenhouse gas emissions, and can be used to produce distributed energy, or energy that is delivered locally at the point of generation, as opposed to relying on a large-scale transmission and distribution system. The primary impediments to the growth of solar PV energy have been price, intermittency, and access to the electricity grid. The price of solar PV has declined precipitously over the past decade, however, making this form of energy increasingly cost competitive.[36]

While solar energy is becoming more cost competitive, there are other important physical and market barriers to consider. First, like wind, solar energy is infinitely renewable but also intermittent. Because the sun only shines for a certain number of hours every day and the intensity of the sunlight available varies based on geography, time of year, and weather, relying on solar power as a consistent form of energy requires developing more efficient and affordable forms of storage. Second, the growth of solar energy as a primary source of electricity requires considering important questions of access to the electricity grid. Like wind, large-scale solar installations are often in remote locations and may require expensive interconnections with, or the construction of new large-scale transmission lines to carry the solar power to the national electricity grid.

Solar energy also raises a distinct set of electricity grid questions. As mentioned, solar energy can be used as an effective form of distributed energy, or energy that is locally produced and used. If, for example, a homeowner installs solar panels on their home, they may be able to produce enough energy to go "off grid", or to retire entirely on their domestic system to meet all of their energy needs. It might also be the case, however, that the domestic solar installation does not meet the homeowner's complete electricity needs, resulting in the homeowner choosing to remain connected to

---

[35] DUVIVIER, *supra* note 17, at 19.

[36] National Renewable Energy Laboratory, *New Report from NREL Documenting Continuing PV and PV-Plus Storage Cost Declines,* (Nov. 12, 2021), https://www.nrel.gov/news/program/2021/new-reports-from-nrel-document-continuing-pv-and-pv-plus-storage-cost-declines.html.

the grid to fill electricity gaps. Or it might be the case that the homeowner is able to produce more electricity than she needs using her solar PV system, so she wants to be connected to the grid in order to sell her excess solar PV generated electricity to the grid for use by other consumers. This relatively new form of interaction between domestic and large-scale electricity producers is called net-metering. As Welton describes the process:

> [T]he "net" aspect of the policy comes from the fact that the transactions are monitored by a single meter, which counts upwards when the consumer is drawing in grid power, and back downwards when the consumer is providing power to the grid. Net metering's popularity is largely due to its simplicity. . . . Net metering is also effective: because it makes investment in solar panels pay off relatively quickly, it has been one of the key policy drivers of the recent solar "boom."[37]

Net-metering, thus, creates a new form of exchange and interaction between individual consumers and large-scale electricity producers that gives the individual consumer more power over their source of power, and more capacity to draw upon the national system both as a back-up source of electricity and as a consumer source for the excess electricity that their home-unit system produces. Because the U.S. electricity grid was "built to handle movement of power from large central power stations to end consumers, and not to manage two-way shipment of power from those consumers," these new forms of interconnection to the grid create a number of challenges, including unpredictability of energy supply, pricing questions, and physical wear and tear challenges.[38] As the number of residential and commercial solar installations grow, many public utilities—traditional electricity suppliers and the entities that continue to maintain ownership and control over much of the large-scale, U.S. electricity transmission and distribution systems—and other opponents of the expansion of solar energy have begun to oppose the growth of distributed energy based on economic and equity arguments, arguing that:

> [C]onsumers with solar panels are "free riding" off the grid: by running the meter backwards, sometimes all the way to zero, net metering allows them to escape from paying their fair share of grid maintenance costs, even though they rely on the grid's services whenever they are under- or over-producing power . . . This "free riding"—or, more accurately, cross-subsidization—appears particularly

---

[37] Shelley Welton, *Clean Electrification*, 88 U. COLO. L. REV. 571, 592 (2017).

[38] DAVIES ET. AL., *supra* note 1, at 154.

> egregious when coupled with statistics showing that predominantly wealthier consumers put solar panels on their roofs. As these consumers enjoy the benefits of self-generation, lower-income consumers who cannot afford solar panels are left shouldering a rising proportion of grid maintenance cost.[39]

The continued expansion of distributed solar power installations, therefore, involves addressing important questions about how to maintain reliable, low-cost electricity for everyone while allowing the growth of renewable energy generation.

Finally, like wind, the aesthetic and environmental implications of both small scale and large-scale solar installations also engender opposition to particular solar projects. Some communities, for example, have banned rooftop solar panels on the basis that they are aesthetically unpleasing and violate home-owners association policies. Larger scale projects, because they tend to be located in open, remote areas of land often must consider the impact of the development and operation of the project on wildlife habitat, including species protected under the Endangered Species Act, such as the desert tortoise.[40] Also, because many of the projects are located on public lands in hot, arid areas, questions often arise about the impact on water use and availability in the area, and the trade-offs inherent in using public lands for the development of solar energy, as opposed to other, potentially conflicting uses.

The final two sources of renewable energy to discuss are biomass/biofuels and geothermal energy. Both of these forms of energy are conventional technologies that have become the subject of greater interest and development in recent years. Biomass is a very broad term that "refers to a vast array of non-fossil biological materials, ranging from firewood to crop residue, and from dung to liquid fuels like ethanol produced from sugarcane, corn, or cellulosic material. . .[or] from algae and methane captured from municipal landfills."[41] Biomass provides one of the most widespread and essential forms of energy in large parts of the world where reliable access to large-scale electricity continues to lag. In recent years, technological advances in this area have focused on using advanced forms of biomass to create cleaner-burning biofuels to replace or offset the consumption of petroleum-based fuels.

The use of biofuels, especially first generation corn ethanol, has raised far-reaching questions about the impacts that biofuel production have on existing patterns of land and water use, with

---

[39] Welton, *supra* note 37, at 594–595.

[40] Defenders of Wildlife v. Zinke, 856 F.3d 1248 (2017).

[41] DAVIES ET. AL., *supra* note 1, at 157.

particular concerns that policies promoting biofuel production, such as the U.S. Renewable Fuel Standard (RFS),[42] could incentivize intensive land use and agricultural shifts from food crops to biofuel crops, with negative consequences for food security worldwide.[43] These concerns have prompted a 'food versus fuel' debate that continues to be the subject of inquiry. That debate has become particularly intense recently as the Ukraine war has cut off access to Ukrainian grains and led to higher food prices. Recognizing both potential trade-offs between food and fuel and the need for more advanced fuels, many ongoing efforts in the biofuel area focus on developing second- and third- generation biofuels that are made using, respectively, non-edible crop parts/non-food crops and algae-based biofuels that reduce competition with food-production and minimize the environmental intensity of the fuel production.

Geothermal energy systems are "simply exploitable concentrations of the Earth's natural heat."[44] Geothermal hot springs have been used by humans for thousands of years, but beginning in the 1960s, these sources have been used more instrumentally to produce electricity, with the United States being the largest global user, but with other countries such as Iceland, El Salvador, and the Philippines also relying on geothermal energy for part of their electricity supply.[45] While geothermal only contributes 0.4 percent of U.S. electricity generation, former U.S. Secretary of Energy, Steven Chu, characterized geothermal energy as "effectively unlimited" and ongoing research suggests that new technologies could expand the opportunities for cultivating geothermal energy sources in particular parts of the United States.

Although it is not, strictly speaking, an energy source, energy efficiency can also be considered as a type of renewable energy, in the sense that it can reduce the amount of energy that is produced from fossil fuels. Although it will not be a focal point of this chapter, energy efficiency can be a promising approach to reducing use of fossil fuels. It has the advantage of directly saving money for energy users, which can partially (and sometimes totally) offset the cost of consumers of implementing efficiency measures. It can also save money for the

[42] For example, the United States has a Renewable Fuel Standard (RFS) that is administered by the EPA under the Clean Air Act and initially mandated that a minimum of 4 billion of renewable fuel be used in the nation's gasoline supply in 2006, with that amount increasing to 7.5 billion gallons by 2012 and, later, as dictated by the Energy Independence and Security Act (EISA), to 9 billion gallons in 2008, and to 36 billion gallons in 2022, with the amount that corn-based ethanol being capped at 15 billion gallons. Alexandra B. Klass & Andrew Heiring, *Life Cycle Analysis and Transportation Energy*, 82 BROOK. L. REV. 485, 494 (2017).

[43] *See generally Carmen Gonzalez, The Environmental Justice Implications of Biofuels*, 20 UCLA J. INT'L L. & FOREIGN AFF. 229 (2016).

[44] DUVIVIER, *supra* note 17, at 220.

[45] DAVIES ET. AL., *supra* note 1, at 155.

system as a whole by eliminating the investment that would otherwise be required in new capacity to meet energy needs.

II. The Electric Power Sector

Each of the sources of energy described above feeds one, or both, of the two most dominant sources of greenhouse gas emissions in the United States—the electric power and transportation sectors. With the electric power sector accounting for 25 percent of emissions, and the transportation sector accounting for 27 percent of emissions,[46] tackling emissions from these two sectors is critical to domestic efforts to address climate change.

The electric power sector rests at the heart of domestic economic activity and is massive in scale, making any efforts to restructure the sector to reduce greenhouse gas emissions physically and legally complex. To contextualize the scale and intensity of the electric power sector, according to the EPA, in 2021:

> [A]bout 4,116 billion kilowatt hours (kWh) (or about 4.12 trillion kWh) of electricity were generated at utility-scale electricity generation facilities in the United States. About 61% of this electricity generation was from fossil fuels—coal, natural gas, petroleum, and other gases. About 19% was from nuclear energy, and about 20% was from renewable energy sources.[47]

While many federal agencies, such as the Department of Energy, the Nuclear Regulatory Commission, the Department of Interior, and the EPA, are involved in regulating different aspects of the energy system, one of the most central players in the energy and, especially, in the electricity contest is the Federal Energy Regulatory Commission (FERC). FERC's stated mission is to "[a]ssist consumers in obtaining reliable, efficient and sustainable energy services at a reasonable cost through appropriate regulatory and market mean"[48] and its extensive responsibilities in the energy context include regulating the interstate transmission of natural gas, oil, and electricity, as well as natural gas and hydropower projects.

FERC's responsibilities are broad and often overlap not only with other federal agencies, but also with state governments. The area of energy regulation, generally, involves a highly complex set of relationships between federal, state, and local governmental oversight, with the level of federal control varying among the

---

[46] U.S. Environmental Protection Agency, *Sources of Greenhouse Gas Emissions*, https://www.epa.gov/ghgemissions/sources-greenhouse-gas-emissions.

[47] *What is U.S. Electricity Generation by Energy Source?*, *supra* note 12.

[48] FERC, Federal Energy Regulatory Commission, *About FERC*, https://www.ferc.gov/about/about.asp.

different sources of energy, with the most federal oversight over nuclear energy, followed by coal (which is heavily regulated by environmental laws), with the least amount of federal oversight and control in the areas of oil and gas law, the regulation of which has historically fallen primarily to the states, with the exception of pipeline development and maintenance, where there is more federal oversight. (Note, however, that there is an important exception: oil and gas are often located on public lands or offshore areas that are controlled by the federal government.) With respect to the electric power system, historically, due to the interstate nature of the majority of transactions, the federal government has exercised significant control over the three key stages of the electric power lifecycle, generation, transmission, and distribution, with state commissions, over time, playing an increasingly important role in regulating the investor-owned utilities (IOUs), that shape the electric power market.

A. Electricity Regulation in Transition

Historically, the electric power system has been owned and operated by a small group of entities called utilities, including both public and private utilities. These entities operate in a vertically integrated fashion, meaning that at one end there is an electric power provider (e.g., a large coal fired power plant) that produces electricity that is then moved along high voltage transmission lines, and then down through smaller scale distribution lines to the hundreds of thousands of consumers that are waiting at the other end. As a result, "[p]roducers and consumers are linked together through transmission and distribution lines," meaning that the customers of these vertically integrated utilities, which includes the majority of Americans, "are literally at the pricing mercy of the producers."[49] These vertically integrated utilities developed because there are extremely high sunk costs involved in creating the infrastructure for producing and selling electric power, including the generation, transmission, and distribution facilities. And, once an entity has invested in the necessary infrastructure, they want to maintain as much control over that system as possible in order to recover their sunk costs over time. In the early days of the development of the electric power sector, only a small handful of entities could afford to invest in this area, meaning that this small handful of actors could exercise monopoly power. Absent governmental intervention, the entity exercising this monopoly power could charge whatever price they wanted for electricity. As a result, very early on, the state government stepped in to regulate these entities in order to ensure that the rates utilities charge for power are "just and reasonable."

---

[49] DAVIES ET. AL., *supra* note 1, at 308.

For instance, such language is found in the Federal Power Act of 1935, one of the pillars of energy law, which addresses wholesale transactions rather than retail sales to consumers. It provides that "[a]ll rates and charges made, demanded, or received by any public utility for or in connection with the transmission or sale of electric energy . . . and all rules and regulations affecting or pertaining to such rates or charges shall be just and reasonable."[50]

As a result, for many years the electric power sector was closely held and controlled by a relatively small group of actors. As Scott details, while "[t]he utility industry includes a variety of participants, from small power producers, to federal power agencies, to power marketers. Within the retail market, however, customers are served primarily by investor-owned utilities ("IOUs"), cooperatives, and publicly-owned utilities ('publics')."[51] Today, with roughly seventy percent of retail customers being served by IOUs, the operation and regulation of these entities is critical to ensuring fair and equitable access to electricity, as well as to accomplishing other policy goals, such as moving towards a cleaner, lower-carbon electric power system. These IOUs are largely regulated by state public utility commissions, who control the extent to which the IOU can price electricity and, thus, the extent to which the IOUs can recover the costs of bringing new generation facilities online, investing in efficiency, improving distribution lines, or otherwise investing in the electric power system. Consequently, "regulatory commissions have significant power to determine how and when the electric utility grid will evolve, the types of generation facilities that will be constructed, and the amount of money and capital investment that will be expended toward various resource options, including renewables and energy efficiency."[52] Equally important, this is a price and profit driven field. As a result, "unless legislation specifically requires public utility commissions to consider environmental, technological, or policy matters, they will focus—almost exclusively—on rate impacts to current customers."[53]

Until the 1970s, this traditional system remained largely static. Beginning in the 1970s, however, driven by the oil crisis, the emerging environmental movement, and economic and political uncertainty, there were growing calls to restructure the electric power market to allow more competition and to create greater room for smaller, cleaner sources of energy like wind power to be brought

---

[50] The Federal Power Act, 16 U.S.C. § 824d(a) (2012).

[51] Inara Scott, *Teaching an Old Dog New Tricks: Adapting Public Utility Commissions to Meet Twenty-First Century Climate Challenges*, 38 Harv. Envtl. L. Rev. 371, 374–75 (2014).

[52] *Id.*

[53] *Id.*

online. The early move toward restructuring the electric power system was driven by the Public Utility Regulatory Policies Act of 1978 (PURPA).[54] PURPA is widely heralded as "inaugurat[ing] the process by which the traditional structure of then the utility system disintegrated."[55] It contained the first congressional move towards cracking the existing monopoly and introducing more competition into the electric power industry. Driven, in part, by competition and environmental concerns, PURPA's key lasting effects were to introduce the element of competition and to create greater room for the growth of renewable and cogeneration facilities by opening up the market to these smaller competitors. The primary way that PURPA promoted renewable energy was by establishing a new class of small generating facilities, called Qualifying Facilities (QFs), that would receive special rate and regulatory treatment.

Although PURPA opened the door to rethinking the electricity industry, it wasn't until the 1990s, with the enactment of the Energy Policy Act of 1992[56] and the issuance of FERC Order 888,[57] that the current period of restructuring began. One of the primary impediments to introducing greater competition into the market and, therefore, creating more opportunities for independent generators to compete is the fact that a small handful of actors control the transmission lines. Without affordable access to privately owned, high voltage transmission lines, independent producers have no way to enter the market. As a result, in order to allow the growth of a more dynamic electricity market, it was necessary to find ways to broaden access to the transmission grid. Both the Energy Policy Act of 1992 and FERC Order 888 broadened affordable access to the transmission grid, creating more opportunities for competition and diversification among energy generators, including the growing renewable energy industry. As a result of these and other efforts, we are in a period of transition with our electric power grid, with many states engaging in complementary efforts to encourage more market competition and expanded capacity for independent energy sources

[54] 16 U.S.C. § 2601 *et seq.* (hereinafter PURPA).

[55] RICHARD F. HIRSH, POWER LOSS: THE ORIGINS OF DEREGULATION AND RESTRUCTURING IN THE AMERICAN ELECTRIC UTILITY SYSTEM 119 (1999).

[56] Energy Policy Act of 1992, 42 U.S.C. §§ 13201–13556 (1997) (In key part, the Act authorized FERC to order individual utilities to provide transmission services to unaffiliated wholesale power generators on a case-by-case basis).

[57] FERC, *Promoting Wholesale Competition Through Open-Access Non-discriminatory Transmission Services by Public Utilities Recovery of Stranded Costs by Public Utilities and Transmitting Utilities*, Order No. 888, 18 CFR Parts 35 and 385 (1996) (ordering the "functional unbundling" of wholesale generation and transmission services). *See also* New York v. FERC, 535 U.S. 1, 11 (2002) (defining functional unbundling as requiring "each utility to state separate rates for wholesale generation, transmission, and ancillary services, and to take transmission of its own wholesale sales and purchases under a single general tariff").

to feed the grid, while also working to ensure grid reliability through continuing close oversight by state and federal regulators.

B. Next Generation Electricity: The Smart Grid and the Growth of Renewable Energy

The ongoing restructuring of the domestic electric power grid is a necessary precursor to efforts to facilitate the growth of renewable energy. As discussed, however, fossil fuels remain the dominant source of energy powering the electric power sector. The existing electric power sector is built on the foundations of fossil fuels because, historically, they have provided a low cost, reliable source of energy. Low cost, reliable energy fueled the growth of America and continues to be viewed as a fundamental component of the American way of life. As a result, any longer-term, more expansive shifts in the electric power sector must be done in ways that ensure that our energy supply remains reliable and affordable.

With respect to reliability, the U.S. electric power grid is a massive, complex system or, as Boyd et al. describes it, the electric power grid is "the largest machine in the United States . . . which connects the dam and the factory and the accelerator to transmission and distribution lines to provide them with power."[58] The national grid, operating through three large regional sub-grids,[59] requires constant oversight and management to make sure that the system provides the right amount of power to the right place at the right time.[60] At the moment, the grid is largely one-sided, meaning it operates with very little customer feedback. The system, however, is gradually moving towards a next generation power grid—a "smart grid"—that will enable two-way flows of information between the energy consumers and the energy generators. Moving towards a system that allows constant, two-way communication will improve efficiency and reliability, while also providing opportunities for greater customer choice and greater capacity to integrate renewable power sources. In addition, the smart grid could enable quicker restoration of electricity after power disturbances, such as those brought on by storms such as Super Storm Sandy, Hurricane Florence, Hurricane Maria, or the 2021 winter storms in Texas. As described in Chapters 2 and 8, climate change is increasing the frequency and intensity of storms, resulting in more human

---

[58] William Boyd, Ann Carlson, & Cara Horowitz, *The Machine at the Center of the Clean Power Plan*, Legal Planet (Sept. 27, 2016), http://legal-planet.org/2016/09/27/the-machine-at-the-center-of-the-clean-power-plan/.

[59] The three grids are: The Western Interconnection, The Eastern Interconnection, The Electricity Reliability Council of Texas Interconnection.

[60] The entities helping operate the grid include electric utilities, Independent System Operators (ISOs), Regional Transmission Organizations (RTOs).

disasters. A smart grid would, ideally, be more resilient in the face of such events.

While moving towards a smart grid is desirable for efficiency, safety, and environmental reasons, the existing electric power grid is large and old, making the transition to a smart grid both slow and expensive. During his campaign for office, President Obama committed to pursuing "a major investment in our national utility grid to enable a tremendous increase in renewable generation and accommodate 21st century energy requirements, such as reliability, smart metering, and distributed storage."[61] Once elected, he accelerated grid modernization projects, including the permitting and construction of electric transmission lines that would help integrate renewable electricity sources into the grid, as well as helping bolster grid resiliency and reliability. In 2009, Congress passed the American Recovery and Reinvestment Act (ARRA)[62], which allocated $4.5 billion to the Department of Energy (DOE) to modernize the electric power grid. The larges program that resulted from this investment is the Smart Grid Investment Grant (SGIG). Through SGIG, DOE and the electricity industry were able to jump start efforts to move towards a smart grid, jointly investing "$8 billion in 99 cost-shared projects involving more than 200 participating electric utilities and other organizations to modernize the electric grid, strengthen cybersecurity, improve interoperability, and collect an unprecedented level of data on smart grid operations and benefits."[63]

During President Obama's two terms, the Administration invested in a number of efforts to jumpstart the move towards a more modern grid, as well as to facilitate complementary energy efficiency efforts to reduce demand on the grid.[64] President Obama's efficiency measures, including updating energy conservation standards for many household and commercial lamps and lighting equipment, setting new efficiency goals for federal buildings, and providing new

---

[61] *Barack Obama's Plan to Make America a Global Energy Leader* 7–8, https://grist.org/wp-content/uploads/2007/10/100707_fact_sheet_energy_speech_final.pdf.

[62] ARRA also "provided $5 billion for low-income weatherization programs (including $1,500 in tax breaks), $4.5 billion to green federal buildings, and $6.3 billion for state and local renewable energy and energy efficiency efforts, which included the $3.2 billion Energy Efficiency and Conservation Block Grant (EECBG) Program." Hari Osofsky, *Diagonal Federalism and Climate Change Implications for the Obama Administration*, 62 ALA. L. REV. 237 (2011).

[63] Energy.gov, *Recovery Act: Smart Grid Investment Grant (SGIG) Program*, https://energy.gov/oe/information-center/recovery-act-smart-grid-investment-grant-sgig-program.

[64] *E.g.*, Chris Mooney, *Obama Just Released the Biggest Energy Efficiency Rule in U.S. history*, WASH. POST (Dec. 17, 2015), https://www.washingtonpost.com/news/energy-environment/wp/2015/12/17/meet-the-biggest-energy-efficiency-rule-the-u-s-has-ever-released/?tid=a_inl&utm_term=.536d01c27c16.

and additional resources to improve building efficiency nationwide[65], received little attention in compared to his efforts to reduce greenhouse gas emissions from stationary energy sources under the Clean Air Act. These efficiency efforts, however, were extensive and are estimated to save billions of dollars each year, as well as reduce carbon emissions.

The Trump Administration did not sustain efforts to modernize the grid. Except for investing in critical electric infrastructure in rural areas,[66] most of the Trump Administration's electricity-focus policies undercut efforts to expand or modernize the grid. In 2019, for example, the Trump Administration eliminated the Smart Grid Advisory Committee, which was the advisory committee that provided input on improving smart grids. Then, in 2020, President Trump issued an Executive Order that blocked the installation of bulk power system equipment sourced from rivals of the United States. This move was designed to improve the security of the grid but was operationalized in such a way as to cause confusion among the utilities. Very little was done during the Trump Administration to advance grid modernization efforts.

President Biden has renewed efforts to expand and modernize the electricity grid.[67] In April 2021, the Biden Administration announced new cooperative efforts between the Department of Energy and the Department of Transportation to use existing rights-of-way to facilitate the siting of new high-voltage electric transmission lines. Later that year, in November 2021, Congress jump-started efforts to advance grid modernization and investments in clean energy when it passed the Infrastructure Investment and Jobs Act (Infrastructure Act).

The $1.2 trillion Infrastructure Act represents a comprehensive bipartisan effort to strengthen and expand U.S. infrastructure. The Infrastructure Act is arguably the most sweeping Congressional investment in clean energy and grid modernization to date. As characterized by the Biden White House, the Act will "strengthen our nation's resilience to extreme weather and climate change while

---

[65] Energy.gov, *Obama Administration Launches New Energy Efficiency Measures* (June 29, 2009), https://energy.gov/articles/obama-administration-launches-new-energy-efficiency-efforts.

[66] U.S. Department of Agriculture, *Trump Administration Invests $1.6 Billion in Rural Electric Infrastructure in 21 States* (June 22, 2020), https://www.usda.gov/media/press-releases/2020/06/22/trump-administration-invests-16-billion-rural-electric.

[67] The White House, *Fact Sheet: Biden Administration Advances Expansion and Modernization of the Electric Grid* (April 27, 2021), https://www.whitehouse.gov/briefing-room/statements-releases/2021/04/27/fact-sheet-biden-administration-advances-expansion-modernization-of-the-electric-grid/.

reducing greenhouse gas emissions, expanding access to clean drinking water, building up a clean power grid, and more."[68]

Although the Infrastructure Act sets out to do much more than advance energy infrastructure, it creates the most comprehensive framework to date for transitioning to a clean energy economy, reducing greenhouse gas emissions, and creating a climate-resilient society. In addition to providing the largest ever federal investment in clean energy transmission and the electric grid ($64 billion), the Act provides significant economic investment in public transit; electric vehicle infrastructure; zero-emission school buses; modernizing physical infrastructure (e.g., ports, airports, freight) to make it more sustainable and resilient; improving community resiliency through investments in weatherization of homes and other efforts to protect against drought, heat, and floods; improving access to clean drinking water; and addressing legacy pollution sites.

Notably the Act advances stalled efforts to improve the domestic power infrastructure, including by building thousands of miles of new, resilient transmission lines to facilitate the expansion of renewable energy. It also creates a new Grid Deployment Authority, invests in research and development for advanced transmission and electricity distribution technologies, and promotes smart grid technologies that deliver flexibility and resilience. The law also gave FERC the power to override state resistance to new transition lines in some circumstances. The Act also directs funds to demonstration projects and research hubs for next generation technologies like advanced nuclear reactors, carbon capture, and clean hydrogen.

Following adoption of the Infrastructure Act, in April 2022, the DOE launched its "Building a Better Grid Initiative"[69]. This Initiative creates the foundation for DOE to deploy more than $20 billion in federal financing tools, including more than $2.5 billion in new Infrastructure Act funds targeted for the development of a Transmission Facilitation Program, $3 billion for the expansion of the Smart Grid Investment Grant Program, and more than $10 billion in grants for States, Tribes, and utilities to enhance grid resilience and prevent power outages.

---

[68] *Fact Sheet: The Bipartisan Infrastructure Deal Boosts Clean Energy Jobs, Strengthens Resilience, and Advances Environmental Justice*, WHITE HOUSE (Nov. 8, 2021), https://www.whitehouse.gov/briefing-room/statements-releases/2021/11/08/fact-sheet-the-bipartisan-infrastructure-deal-boosts-clean-energy-jobs-strengthens-resilience-and-advances-environmental-justice/.

[69] Department of Energy, *Biden Administration Launches $2.3 Billion Program to Strengthen and Modernize America's Power Grid* (April 27, 2022), https://www.energy.gov/articles/biden-administration-launches-23-billion-program-strengthen-and-modernize-americas-power.

At the time of writing, many of the DOE projects and the other initiatives funded through the Infrastructure Act are still in the early phases of development, but the Infrastructure Act creates the thickest legal and financial foundation, to date, for expanding and modernizing the grid and advancing clean energy deployment.

Beyond the grid, the federal government also has significant control over energy production because of its ownership of public lands (roughly one-third the area of the U.S.) and control of offshore activities. There are enormous coal, oil, and gas resources in these locations. The legal structure for regulating public lands is complex, and space does not allow us to go into depth. In general, the applicable laws can be divided into three categories. The first consists of laws applying specifically to public lands and providing procedures and standards for private use of those lands. For instance, there is an elaborate set of procedures for issuing offshore oil leases, and these laws may include requirements for consideration of environmental factors. The second set of laws allows some areas to be withdrawn from development. These laws encompass national parks, wilderness areas, and national monuments. The third set consists of environmental laws of general application, requiring environmental impact statements, protecting endangered species, and regulating pollution.

The Obama Administration tended to apply these laws so as to limit use of fossil fuel resources on public lands and offshore. In contrast, the Trump Administration's goal was to encourage American fossil fuel production, and it immediately embarked on a campaign to expand production on public lands and in offshore waters. The five-year plan the Trump Administration proposed for offshore drilling, for example, sought to open up approximately 90% of the Outer Continental Shelf off the U.S. coast to oil and gas exploration as well as to lift a ban on drilling in the Arctic Ocean. This proposal represented the largest single attempted expansion of offshore drilling activity ever proposed.[70]

Legal challenges limited the Trump Administration's ability to fully implement the proposed strategy. The 9th Circuit for example, upheld a lower court decision denying President Trump's efforts to revoke an Obama-era designation of certain areas of the Arctic as withdrawn from leasing,[71] while legal challenges from environmental organizations slowed down the full-scale deployment of the proposed off-shore drilling program in the Gulf of Mexico. The Trump

---

[70] Bureau of Ocean Energy Management (BOEM), *2019–2024 National Outer Continental Shelf Oil and Gas Leasing Draft Proposed Program* (January 2018), https://www.boem.gov/sites/default/files/oil-and-gas-energy-program/Leasing/Five-Year-Program/2019-2024/DPP/NP-Draft-Proposed-Program-2019-2024.pdf.

[71] Center for Biological Diversity v. Bernhardt, 982 F.3d 723 (9th Cir. 2020).

Administration similarly sought to expand drilling on public lands, including through controversial decisions to shrink National Monuments such as Bears Ears and the Grand Staircase-Escalante. By shrinking the size of the Monuments, President Trump sought to expand the public land areas open to oil and gas leasing.

Within hours of President Trump issuing the proclamation to shrink Bears Ears five Native American tribes filed a lawsuit in federal court in Washington, D.C., challenging the president's action. Other legal actions followed. While these cases were still pending, President Biden took office and, on his first day in office, he issued an executive order initiating a review of President Trump's monument rollbacks. Subsequently, in October 2021, President Biden issued a new proclamation that restored Bears Ears to the previous boundaries established by President Obama in 2016. Despite these challenges, offshore drilling and drilling on public lands was substantially expanded during the Trump Administration.

In contrast to the Trump Administration, President Biden promised no more drilling on federal land during his presidential campaign. After coming into office, in January 2021, Biden issued an executive order halting all new oil and gas leases on public lands and in offshore waters and directing the Department of Interior to undertake a review of the exiting federal oil and gas leasing program.[72] This ban was short lived. In June 2021, a federal judge in Louisiana blocked the Biden Administration's temporary ban on new oil and gas leases on public lands and in offshore waters.[73] As a result, the federal government renewed its leasing program and, during President Biden's first year in office, federal agencies approved 3,557 permits for oil and gas leases. This figure exceeded the number of permits the Trump Administration approved in its first year by almost 35%. At this writing, however, the pace of permitting had dropped significantly.

C. State Level Efforts on Renewable Energy

Under the Obama and Biden Administrations, the federal government has taken various steps to move the country towards a cleaner and more efficient energy system. Over the past 15 years, states have similarly begun to take more active steps to improving the efficiency and environmental footprint of the electricity system. One of the primary tools states have used to encourage the growth of renewable energy sources is the Renewable Portfolio Standard (RPS). An RPS requires electric utilities to ensure that a certain percentage of the retail power they sell comes from designated

---

72 Tackling the Climate Crisis at Home and Abroad, 86 FR 7619.

73 Louisiana v. Biden, 543 F. Supp. 3d 388 (W.D. La. 2021).

categories of renewable energy sources. As a result, the goal is to use the RPS to "harness the power of markets to allow participants to find the most efficient result themselves."[74] Each state RPS specifies what power sources qualify as renewable or "alternative," with most states including energy produced from solar, wind, hydropower, and geothermal while others include other sources such as energy from landfill gas recovery, cogeneration, and fuel cells. The number of state RPS laws has grown rapidly. There are now 20 states that have adopted RPS, nine states that have adopted Clean Energy Standards (CES) that requires a certain percentage of a utility's electricity to come from low- or zero-carbon emitting energy sources, two states that have adopted slightly broader Alternative Energy Portfolio Standards, and another eight states that have a voluntary RPS/Alternative Energy Portfolio Standard.

In addition to the basic RPS, CES and Alternative Energy Portfolio Standard requirements, a number of states add "carve-out" provisions to their RPS that specify that a certain percentage of the renewable or alternative energy that is generated must come from a particular source of energy, generally solar energy. These provisions are designed to ensure that there is state-wide investment in particular types of renewable energy, particularly solar energy, which despite being less cost-effective than wind at the moment, is expected to be an important growth industry. New Jersey for example has a RPS that requires 50% of electricity sales in the state to come from renewable sources by 2030, with at least 2.21% of that coming from solar power by 2020, and at least 3,500 MW of that coming from offshore wind by 2030. A number of states also have provisions requiring that a certain percentage of the renewable, or alternative energy, being used to satisfy the portfolio standard be generated in state.

As RPS standards have become more popular and more widely used, they have also given rise to questions of constitutional consistency under dormant commerce clause doctrine, as will be discussed in greater depth in Chapter 7. The U.S. Constitution grants power to Congress "[t]or regulate Commerce. . .among the several states,"[75] thus expressly granting Congress the authority to regulate interstate commerce. The Court has also found a negative, or 'dormant,' authority inherent in the commerce clause that restricts states from "unjustifiably. . .discriminating against or burdening the interstate flow of commerce."[76] The dormant commerce clause primarily operates to prevent state. In the context of an RPS, while,

[74] Lincoln Davies, *Power Forward: The Argument for a Federal RPS*, 42 CONN. L. REV. 1339, 1357 (2010).

[75] U.S. Const. art.1 § 8, cl.3.

[76] Baldwin v. G.A.F. Seelig, Inc., 294 U.S. 511, 522 (1935).

as Klass and Rossi note, "there is nothing constitutionally suspect about encouraging or even requiring renewable energy, some particular aspects of RPS standards have raised dormant commerce clause concerns."[77] Certain elements of state RPS and Alternative Energy Portfolio Standards, however, raise greater legal concerns. For example, Ohio, in common with other states, requires that half of the renewable energy capacity be generated in state. Similarly, other states have some have formulas or subsidies that favor in-state sources. The RPS elements that favor in-state sources or otherwise differentiate between in-state and out-of-state sources have given rise to a number of challenges and are likely to do so in the future. To date, however, the core components of state RPS standards have withstood judicial scrutiny.

Although many states have taken steps to increase renewable, and alternative energy generation and use, California is an example of a state that has taken a variety of different, increasingly progressive steps to encourage a more comprehensive overhaul of their state energy system. In 2006, then Governor Schwarzenegger signed Senate Bill 107, requiring California's three major utilities to deliver at least 20 percent of their electricity from renewable sources by 2010. Then, in 2009, Gov. Schwarzenegger directed California's Air Resources Board (CARB) to adopt regulations increasing California's RPS to 33 percent by 2020 and expanded the RPS to apply not just to the three major utilities, but also to all load serving entities, including investor-owned utilities, publicly owned utilities, direct access providers and community choice aggregators. In 2011, Gov. Jerry Brown codified the 33 percent by 2020 target before expanding the target again in 2015 to require retail sellers and publicly owned utilities to procure 50 percent of their electricity from eligible renewable energy resources by 2030. As a result, by 2015 California had adopted an aggressive goal requiring, by 2030, 50 percent of electricity to come from renewable sources.

Then, in 2015, California passed the Clean Energy and Pollution Reduction Act (Senate Bill 350), which set ambitious new clean energy, clean air and greenhouse gas emissions reduction goals, namely reducing greenhouse gases to 40 percent below 1990 levels by 2030 and to 80 percent below 1990 levels by 2050. In addition, SB 350 increased California's renewable electricity procurement goal from 33 percent by 2020 to 50 percent by 2030 and required the state to double statewide energy efficiency savings in electricity and

---

[77] Alexandra B. Klass & Jim Rossi, *Revitalizing Dormant Commerce Clause Review for Interstate Coordination*, 100 MINN. L. REV. 129, 165 (2015).

natural gas end uses by 2030.[78] Among the other ambitious changes SB 350 mandates, it also directs state agencies to undertake various studies to identify and assess:

- Barriers to, and opportunities for, solar photovoltaic energy generation.
- Barriers to, and opportunities for, access to other renewable energy by low-income customers.
- Barriers to contracting opportunities for local small businesses in disadvantaged communities.
- Barriers for low-income customers to energy efficiency and weatherization investments, including those in disadvantaged communities.
- Recommendations on how to increase access to energy efficiency and weatherization investments to low-income customers. The Energy Commission conducted the Low-Income Barriers Study, Part A—Adopted, which was adopted in 2016.
- Barriers for low-income customers to zero-emission and near-zero-emission transportation options, including those in disadvantaged communities.
- Recommendations on how to increase access to zero-emission and near-zero-emission transportation options to low-income customers, including those in disadvantaged communities. The California Air Resources Board (CARB) conducted the study in consultation with the Energy Commission. The Low-Income Barriers Study, Part B: Overcoming Barriers to Clean Transportation Access for Low-Income Residents—Final Guidance Document was released in February 2018.[79]

The emphasis on low-income communities is particularly notable. It reflects a trend by state legislatures to respond to environmental justice concerns in crafting new energy and climate legislation.

California has adopted a number of strategies to facilitate compliance with its ambitious 50% RPS, including measures aimed to improve energy storage capacity, to enable smaller renewable power generators to come online, to expand solar energy capacity, and to encourage distributed energy through net-metering programs.

---

[78] California Energy Commission, *Clean Energy and Pollution Reduction Act—SB 350*, https://www.energy.ca.gov/rules-and-regulations/energy-suppliers-reporting/clean-energy-and-pollution-reduction-act-sb-350.

[79] *Id.*

Concerning storage, since 2013, California has imposed an energy storage mandate on its utilities that requires the state's three IOUs to develop a significant amount of distributed energy storage resources. Energy storage capacity, as discussed, is essential to bringing more renewable energy power online becomes it minimizes the disruption to the grid that can be caused by intermittency.

In order to enable smaller renewable power generators to come on-line, California uses a feed-in tariff (FIT), a form of price guarantee that was pioneered and used widely in Germany to promote renewable energy. FITs work by guaranteeing certain generators—generally small-scale renewable power generators—access to the grid (feed-in) as well as requiring the utilities to buy the power that these generators produce at an above-market price (tariff) for a set period of time. In this way, the FIT allows small generators certainty and encourages investment in renewable power. California's FIT, the Renewable Market Adjusting Tariff (ReMAT), enables smaller renewable power generators to enter into a fixed-price standard contract for ten, fifteen, or twenty years, to export electricity to California's three large IOUs.[80]

Finally, customers who install small, solar, wind, biogas and fuel cell generation facilities of up to 1 megawatt capacity are eligible for California's net-metering program, Net Energy Metering (NEM). NEM allows customers who generate more electricity than they need to receive a financial credit for the power that they feed back to the utility; the financial credit is then used to offset the customer's electricity bill. The California Public Utility Commission is in the process of reconsidering its net metering program to further incentivize the installation of customer-cited renewable energy and to advance efforts to meet California's ambitious climate goals even faster.

Together with a number of other complementary initiatives, these steps help California facilitate a long-term shift towards a more diversified electric power portfolio that includes renewable energy as a core component of the system. California and other states have also taken steps to encourage energy efficiency. California's Public Utility Commission began pursuing energy conservation as a goal in the 1970s, and it is probably no coincidence that its per capita electricity use leveled off while levels continued to rise steadily at the national level. Utility firms, like other businesses, make money selling their product. Thus, it requires carefully designed incentives to induce them to encourage reduced use by their customers. Other strategies

---

[80] California Public Utilities Commission, *Renewable Feed-In Tariff (FIT) Program*, http://www.cpuc.ca.gov/feedintariff/. *See also* Felix Mormann, Dan Reicher, Victor Hanna, *A Tale of Three Markets: Comparing the Renewable Energy Experiences of California, Texas, and Germany*, 35 STAN. ENVTL. L.J. 55, 79–80 (2016).

for improved energy efficiency include changes in building codes and subsidies for improved insulation and weatherproofing of residences.

With the national electric power system in flux, California's efforts to create the foundations for a more diversified and flexible system create a helpful blueprint for future state and national efforts to restructure the domestic electric power system.

III. The Transportation Sector

The transportation sector accounted for 27 percent of U.S. greenhouse gas emissions in 2020, exceeding the electric power sector in greenhouse gas emissions by 2 percent. Greenhouse gas emissions from this sector come from the burning of fossil fuels, including gasoline and diesel, in automobiles, trucks, planes, trains, and boats. With light duty vehicles contributing 61 percent, and medium and heavy-duty vehicles contributing another 23 percent of emissions, cars and trucks constitute the dominant source of domestic transportation emission.[81] As a result, early efforts to reduce emissions from the transportation sector focused on controlling the fuel consumption of vehicles through fuel economy standards and vehicle tailpipe emission standards. In common with the electric power sector, however, the transportation sector is large and unwieldy and involves varying levels of control by the federal, state, and local authorities, making efforts to reduce emissions across the sector challenging. As Arroyo et al. emphasize, tackling "transportation-sector emissions can only be significantly reduced by using policy levers at different levels of government and attacking all factors underlying transportation emissions (including the fuel consumption of vehicles, the carbon content of fuels, and the amount of travel that occurs)."[82]

The vast majority of the domestic vehicle fleet runs on fossil fuels. As a result, tackling emissions from the transportation sector requires limiting how much fuel the vehicle fleet consumes, how carbon intensive that fuel is, and how many miles vehicles travel each year. The first element of this strategy—controlling vehicle fuel consumption—has been the focus of federal regulation since 1975, when Congress established the first fuel economy standards for new passenger cars, largely driven by concerns about dependency on foreign oil following the 1973 Arab Oil Embargo. Since that time, CAFE standards have been used as a primary tool for controlling automobile fuel consumption.

---

[81] U.S. EPA, *Fast Facts on Transportation Greenhouse Gas Emissions*, https://www.epa.gov/greenvehicles/fast-facts-transportation-greenhouse-gas-emissions.

[82] Vicki Arroyo et al., *State Innovation on Climate Change: Reducing Emissions from Key Sectors While Preparing for a "New Normal"*, 10 HARV. L. & POL'Y REV. 385, 389 (2016).

As discussed further in Chapter 7, with an exception that applies to California, the Clean Air Act preempts states from regulating emissions from new motor vehicles or new motor vehicle emissions, making this an area where federal efforts are paramount. As a result of frustration with the slow pace of federal efforts to limit greenhouse gas emissions from automobiles, in 1999, a small private organization filed a petition requesting that EPA begin regulating four greenhouse gases, including carbon dioxide, emitted from new motor vehicles. As examined in greater detail in Chapter 6, this petition led to the decision in *Massachusetts v. EPA*[83] and, subsequently, in 2009, to EPA issuing the section 202(a)(1) endangerment finding that triggered an obligation on EPA's part to begin regulating greenhouse gas emissions from new automobiles.[84] That same year, EPA also granted a waiver of Clean Air Act preemption to California for its greenhouse gas emission standards for motor vehicles beginning.[85]

Following these two path-breaking decisions, in May 2009, an agreement between EPA, the Department of Transportation (DOT), state regulators, and the auto industry established the first-ever national wide greenhouse gas emission standards for light-duty vehicles and the most progressive fuel efficiency improvements in 30 years. The 2009 rule applied to model years 2012–2016. This rule was followed, in 2012, by another rule requiring additional reductions in greenhouse gas emissions and additional improvements in fuel economy for light-duty vehicles for model years 2017–2025.

In 2014 and 2015, EPA also finalized gasoline standards that further contribute to vehicle efficiency, for passenger cars, light-duty trucks, medium-duty passenger vehicles, and some heavy-duty vehicles. As a result of these rules, a model year 2025 vehicle is expected to emit one-half of the GHGs that a model year 2010 vehicles emitted.

During the era of the Obama Administration, the EPA also took a number of other steps to address emissions from automobiles, including creating a Green Vehicle Guide that provides greenhouse gas ratings for all new automobiles, and taking steps to incentivize the growth of electric vehicles by pledging $2.4 billion in federal

---

[83] 549 U.S. 497 (2007).

[84] 42 U.S.C. § 7521(a)(1).

[85] U.S. EPA, *Timeline of Major Accomplishments in Transportation, Air Pollution, and Climate Change*, https://www.epa.gov/transportation-air-pollution-and-climate-change/timeline-major-accomplishments-transportation-air.

grants to support the development of next-generation electric vehicles and batteries.[86]

Efforts during the Obama Administration to tackle greenhouse gas emissions from vehicles were significant and created a pathway toward a significantly smaller greenhouse gas footprint for the vehicle industry. President Trump, however, almost immediately began the process of rolling back emissions limitations and federal fuel economy standards for automobiles. In August 2018, the EPA released a proposed rule—the Safe Affordable Fuel Efficient (SAFE) Vehicles proposal—that would freeze emissions and fuel-efficiency standards for cars after 2021 and would revoke the waiver of CAA preemption the EPA granted California to establish its greenhouse gas emissions standards. Subsequently, in September 2019, the U.S. Department of Transportation's National Highway Traffic Safety Administration (NHTSA) and EPA issued a final action entitled the "One National Program Rule" to enable the federal government to provide nationwide uniform fuel economy and greenhouse gas emission standards for automobile and light duty trucks.[87] This rule finalized parts of the earlier proposed SAFE proposal sought to preempt state and local tailpipe GHG emissions standards. It also revoked California's waiver under the CAA to regulate GHG emissions standards from automobiles.

On his first day in office, President Biden issued an Executive Order directing EPA to consider whether to revise, rescind, or suspend both parts of the SAFE rule. In August 2021, EPA proposed new fuel efficiency standards for model years 2023 to 2026. The final rule was published on December 30, 2021 and went into effect on February 28, 2022. The Biden Administration's rule reinstates ambitious greenhouse gas emissions standards for passenger cars and light trucks through model year 2026. In addition, in March 2022, the EPA reinstated California's tailpipe waiver.

Despite federal regulatory flux level, as will be discussed in Chapter 7, a number of states, led by California, have played important leadership roles in efforts to reduce the carbon footprint of the domestic transportation sector. Because the CAA limits the ability of state to impose additional pollution controls on automobiles, state efforts have focused on promoting low carbon alternatives to traditional, fossil-fuel vehicles, and minimizing the

---

[86] Energy.gov, *President Obama Announces $2.4 Billion in Grants to Accelerate the Manufacturing and Deployment of Next Generation of U.S. Batteries and Electric Vehicles* (Aug. 5, 2009), https://energy.gov/articles/president-obama-announces-24-billion-grants-accelerate-manufacturing-and-deployment-next.

[87] The Safer Affordable Fuel-Efficient (SAFE) Vehicles Rule Part One: One National Program, 84 Fed. Reg. 51,310.

number of miles driven through smarter land-use planning and improved access to public transportation.

States have taken a leadership role in reducing transportation emissions in a number of ways, including both state-based and regional efforts. At the regional level, for example, eleven northeast and mid-Atlantic States, and the District of Columbia have partnered to create the Northeast Electric Vehicle Network of the Transportation and Climate Initiative, through which the relevant state agencies are cooperating to create the groundwork for a regional electric vehicle (EV) system and "engaging in important planning work to remove barriers to the widespread adoption of electric vehicles and ensure that public charging stations are placed in strategic locations that both maximize usage and facilitate interstate travel."[88] Similarly, Oregon, Washington, and California are partnering through the West Coast Electric Highway program to create the infrastructure to enable EV use up and down the West Coast.

At the state level, complementing its ambitious RPS, California has implemented a Zero Emission Vehicle (ZEV) standard, pursuant to its authority under the Clean Air Act.[89] To date, 14 other states have adopted California's ZEV standard,[90] with the governors of seven of these states also partnering to try to put 3.3 million zero-emission vehicles on the road by 2025, and joining the International ZEV Alliance, where they work together with international partners to promote the growth of ZEV vehicles.[91]

Despite advances with respect to ZEVs, the national charging infrastructure needed to create an effective system is lacking, making this an area of critical investment need in the coming years. Recognizing this gap, in the Infrastructure Act, Congress designated $5 billion in investment in state-administered grants for deploying electric vehicle charging stations nationwide. And, in February 2022, the Biden Administration announced the Departments of Transportation and Energy would be working together over the next five years to build out a national electric vehicle charging network.

---

[88] Transportation and Climate Initiative of the Northeast and Mid-Atlantic States, *The Northeast Electric Vehicle Network*, http://www.transportationandclimate.org/node/30.

[89] Cal. Code Regs. tit. 13, § 1962 (2015). *See also*, Arroyo et al., *supra* note 82, at 390–91.

[90] California Air Resources Board (CARB), *States that Have Adopted California's Vehicle Standards Under Section 177 of the Federal Clean Air Act*, (May 13, 2022), https://ww2.arb.ca.gov/sites/default/files/2022-05/§177_states_05132022_NADA_sales_r2_ac.pdf.?

[91] Arroyo et al., *supra* note 82, at 390–91.

As discussed in greater detail in Chapters 6 and 7, beginning in 2006, California also piloted efforts to develop a state-level low carbon fuel standard (LCFS). The California Global Warming Solutions Act, or AB 32,[92] provided broad authority to CARB to develop regulations implementing the Act' far-reaching climate goals. Subsequently, in 2007, by executive order,[93] Governor Schwarzenegger established a "statewide goal" of reducing the carbon intensity of California's transportation fuels by at least 10% by 2020 and called on CARB to establish a LCFS to help achieve this goal.

Pursuant to this mandate, effective 2010, CARB developed the LCFS regulatory program.[94] The regulations establish a baseline of the average carbon intensity of vehicular fuels consumed in California and then mandate that suppliers of vehicular transportation fuels reduce their average carbon intensity from that baseline by set amounts between 2011 and 2020.

While California's LCFS has evolved in response to litigation, in 2015, CARB re-adopted the LCFS regulation[95] to address procedural issues. The standard was then amended in 2018 to strengthen the carbon intensity benchmarks. And, as will be discussed further in Chapter 7, in 2014, the Ninth Circuit rejected a discrimination claim against California's LCFS, clearing the pathway towards full implementation.[96] Oregon and Washington State are adopting similar standards.

In addition to playing leadership roles in the areas of low emissions vehicles and low carbon fuel standards, states also play pivotal roles in influencing land-use planning decisions, development patterns, and modes of mobility that influence the degree to which driving is necessary. As Arroyo et al., emphasize, states play an important role in determining whether there are "[a]ttractive alternatives to driving single-occupancy vehicles."[97] States can shape automobile use and general mobility patterns by focusing on improving public transit systems, incentivizing carpooling, encouraging smart city planning that focuses on high-density, mixed-

---

[92] AB 32 (Nunez), Chapter 488, California Statutes of 2006, codified at CAL. HEALTH & SAFETY CODE § 38500 *et seq.*

[93] State of California, Office of the Governor, Executive Order S-01-0 (Jan. 18, 2007).

[94] California Air Resources Board (CARB), *Low Carbon Fuel Standard*, https://ww2.arb.ca.gov/sites/default/files/2020-09/basics-notes.pdf.

[95] *See Notice of Decision, Re-Adoption of the Low Carbon Fuel Standard*, CAL. AIR RES. BD. (Oct. 2, 2015), http://www.arb.ca.gov/regact/2015/lcfs2015/nodlcfs.pdf.

[96] 730 F.3d 1070, 1077 (9th Cir. 2013) cert. denied, 573 U.S. 946 (2014).

[97] Arroyo et al., *supra* note 82, at 393.

use developments, and investing in "complete street" programs that facilitate mixed uses, such as bicycling and walking.[98]

Sticking with California, as an example of state experimentation with tools to limit carbon emissions economy-wide, California's Sustainable Communities and Climate Protection Act of 2008[99] creates a platform for coordinating transportation and land use planning in order to build more sustainable communities. Pursuant to the Act, CARB establishes regional GHG emissions reductions from passenger vehicle targets for 2020 and 2035. Once these targets are established, each of California's Metropolitan Planning Organizations (MPOs) is tasked with preparing a "sustainable communities strategy" (SCS) as a part of its Regional Transportation Plan (RTP) to meet its emissions reduction obligations. As described by CARB, the SCS should "contai[n] land use, housing, and transportation strategies that, if implemented, would allow the region to meet its GHG emission reduction targets," and "[o]nce adopted by the MPO, the RTP/SCS guides the transportation policies and investments for the region." CARB regularly reviews the SCS to ensure that they will enable compliance with the regional emissions reduction targets.[100]

Beyond California, many cities and states are beginning to think more structurally about how to create more climate friendly, resilient cities. As discussed in Chapter 8, many of these efforts are driven not only by concerns about mitigating climate change, but also by a desire to limit the negative effects of climate change. To the extent that these efforts provide both mitigation and adaptation benefits, they are likely to be more widely accepted and more sustainable, in the long-term.

While most of these planning decisions fall to local, regional, and state entities, the federal government can play an important facilitative role by providing resources and incentives to develop more carbon friendly cities and transit systems.[101]

Domestic efforts to address emissions from the transportation sector focus on light, medium, and heavy-duty vehicle use, but another sector of the transportation system that has been subject to recent scrutiny is aviation. Although aviation emissions currently only constitute 2.5% percent of global greenhouse gas emissions, the

---

[98] *Id.*

[99] S.B. 375, 2007 Leg., Reg. Sess. (Cal. 2008).

[100] The targets were most recently updated in 2018. CARB, *Sustainable Communities and Climate Protection Program*, https://ww2.arb.ca.gov/our-work/programs/sustainable-communities-climate-protection-program/about.

[101] *E.g.*, Fixing America's Surface Transportation Act, Pub. L. No. 114-94, 129 Stat. 1312 (2015) (federal transportation legislation that provides grants to states and local governments for transportation investments).

aviation sector is one of the fastest-growing sources of GHGs, with the International Civil Aviation Organization (ICAO) further forecasting that by 2050 aviation emissions could triple.[102] As a result, the international community has directed increased attention to this area. Following efforts by the European Union to begin regulating the greenhouse gas emissions from any flight that flew into, or out of any European Union country under its Emissions Trading Scheme,[103] the United Nations' International Civil Aviation Organization (ICAO) stepped in to mediate an internationally agreed upon approach to controlling greenhouse gas emissions from aircraft. In 2017, following extensive debate, ICAO adopted the first international carbon dioxide standards for aircraft.[104] The new carbon dioxide standards will apply to new aircraft designs, beginning in 2020, and will vary according to the type of aircraft. At the domestic level, in 2016, EPA "finalized a determination under the Clean Air Act that greenhouse gas (GHG) emissions from certain types of aircraft engines contribute to the pollution that causes climate change and endangers Americans' health and the environment."[105] This finding, in common with the endangerment finding under CAA section 202(a), triggers an obligation to establish emissions standards for aircraft engines. And, in 2021 EPA finalized greenhouse gas emission standards for airplanes used in commercial aviation and for large business jets. This EPA standards align U.S. standards with the emissions standards set by the International Civil Aviation Organization (ICAO). At the time of writing, the new aviation standards are subject to litigation.

In common with aviation, greenhouse gas emissions from shipping are on the rise, and absent mitigation efforts, are expected to more than double by 2050.[106] In the most significant international effort to date to begin to curb the growth in emissions from the

---

[102] European Commission, *Reducing Emissions from Aviation,* https://ec.europa.eu/clima/eu-action/transport-emissions/reducing-emissions-aviation_en.

[103] *See generally* Sanja Bogojević, *Legalising Environmental Leadership: A Comment on the CJEU's Ruling in C-366/10 on the Inclusion of Aviation in the EU Emissions Trading Scheme,* 24 J. OF ENVTL. L. 345 (2012).

[104] ICAO, *Historic Agreement Reached to Mitigate International Aviation Emissions,* (Oct. 6, 2016) https://www.icao.int/Newsroom/Pages/Historic-agreement-reached-to-mitigate-international-aviation-emissions.aspx.

[105] U.S. EPA, *EPA Determines that Aircraft Emissions Contribute to Climate Change Endangering Public Health and the Environment,* (July 25, 2016), https://archive.epa.gov/epa/newsreleases/epa-determines-aircraft-emissions-contribute-climate-change-endangering-public-health.html.

[106] *See, e.g.,* European Commission, *Time for International Action on CO2 Emissions from Shipping* (2013), https://climate.ec.europa.eu/system/files/2016-11/marine_transport_en.pdf (noting that: "While shipping is in most cases more fuel-efficient than other transport sectors, its greenhouse gas emissions are substantial and growing fast. Without action, these emissions are expected to more than double by 2050, due to anticipated growth in the world economy and associated transport demand.")

shipping industry, the International Maritime Organization (IMO), the United Nations specialized agency with responsibility for the safety and security of shipping and the prevention of marine pollution by ships, adopted new regulatory requirements in 2016 designed to gather data on ship fuel use. Pursuant to the new IMO requirements, ships of 5,000 gross tonnage and above must collect consumption data for each type of fuel oil they use, as well as additional data the IMO can use to better understand the greenhouse gas profile of the industry. In adopting these new rules, the IMO stressed that the data collected "will provide a firm basis on which future decisions on additional measures, over and above those already adopted by IMO, can be made."[107]

Although these recent steps by the ICAO and IMO represent important steps forward to control emissions from two of the fastest growing areas of transportation, decarbonizing the transport sector remains a serious, unmapped challenge. Mitigation efforts in the transportation sector continue to lag, with transportation emissions growing both domestically and at the international level. Improved efforts to reduce emissions from the transportation sector will require continuing and enhanced efforts at the local, state, national, and international level to reduce fuel consumption from all forms of transportation, reduce the carbon content of all transport fuels, and minimize the total amount of travel that takes place.

IV. Conclusion

This Chapter has mapped some of the advances and continuing challenges that characterize efforts to decarbonize the electric power and transportation sectors. Over the past decade, significant progress has been made towards thinking through viable pathways towards decarbonization in these two areas. We are, however, still in the early stages of developing and implementing such strategies. Former U.S. President Barack Obama prioritized efforts to address climate change using a wide variety of tools. His efforts in this regard were expansive, but inherently incremental since these are processes that will take time and require continuing effort and investment. President Trump stalled and reversed almost all of President Obama's key initiatives and, although the Biden Administration is now in the process of reinvigorating efforts to advance the move to a clean energy economy, these initiatives remain in their early stages at the time of writing.

---

[107] International Maritime Organization (IMO), *New Requirements for International Shipping as UN Body Continues to Address Greenhouse Gas Emissions* (Oct. 28, 2016), http://www.imo.org/en/MediaCentre/PressBriefings/Pages/28-MEPC-data-collection--.aspx.

Further Readings

Vicki Arroyo et al., *State Innovation on Climate Change: Reducing Emissions from Key Sectors While Preparing for a "New Normal"*, 10 HARV. L. & POL'Y REV. 385 (2016).

Marcilynn A. Burke, *Streamlining or Steamrolling: Oil and Gas Leasing Reform on Federal Public Lands in the Trump Administration*, 91 U. COLO. L. REV. 453, 454 (2020).

K.K. DUVIVIER, THE RENEWABLE ENERGY READER (2011).

Alexandra B. Klass & Elizabeth J. Wilson, *Interstate Transmission Challenges for Renewable Energy: A Federalism Mismatch*, 65 VAND. L. REV. 1801 (2012).

Alexandra Klass, Joshua Macey, Shelley Welton, Hannah Wiseman, *Grid Reliability Through Clean Energy*, 74 STAN. L. REV. 969 (2022).

Hari Osofsky, *Diagonal Federalism and Climate Change Implications for the Obama Administration*, 62 ALA. L. REV. 237 (2011).

Shelley Welton, *Clean Electrification*, 88 U. COLO. L. REV. 571 (2017).

Chapter 6

FEDERAL CLIMATE REGULATION

This chapter will examine the federal government's efforts to reduce carbon emissions from energy, using some of the tools discussed previously in this book. We will focus primarily on the actions of the Environmental Protection Agency (EPA), particularly its efforts to reduce emissions from existing sources using the Clean Air Act. However, we will also look at the role of the courts in the area of climate change and at some other relevant areas of federal law. Sharp changes in the political landscape have had a particularly large impact on federal climate policy because of the highly polarized positions of the two parties.

EPA's leading role in setting federal climate policy was somewhat unexpected. With the election of President Obama and Democratic majorities in both Houses of Congress in 2008, many observers expected Congress to pass major climate change legislation, especially because some key Republicans like John McCain had previously supported such legislation. In 2009, the House actually did pass the Waxman-Markey bill, which would have established a national cap-and-trade system. But the bill stalled in the Senate. Just one year later, in 2010, any chance for new legislation died when the Republicans took control of the House. Thus, Congress—the major engine for large-scale policy development—was taken out of the game and remained largely out of the game throughout the Trump Administration and into the early years of the Biden Administration.

The apparent demise of federal legislation as a realistic option for addressing climate change left EPA as the primary forum for moving forward at the federal level. As discussed earlier, EPA had the opportunity to act because (ironically) it had previously lost a case in the Supreme Court in which it tried to disclaim authority over greenhouse gases under the Clean Air Act. In the ensuing years, EPA has worked valiantly to adapt the Clean Air Act to the problem of climate change. But because the statute was primarily designed with conventional air pollutants in mind, it has taken considerable ingenuity on EPA's part to find statutory anchors for its programs. At every turn, the Obama EPA was plagued with legal challenges, and it subsequently remained largely out of the game throughout the Trump Administration.

Notably, political gridlock created both the need for EPA to regulate and the opportunity for it to do so. Gridlock prevented the passage of climate legislation in the early years of the Obama Administration. But after Republicans took control of both Houses of Congress, gridlock also prevented them from overriding EPA's efforts.

Congress has taken some actions to regulate non-CO_2 pollutants, In 2020, the American Innovation and Manufacturing (AIM) Act contained provisions that phase down the use of hydrofluorocarbons (HFCs), which are super-potent greenhouse gases. In 2022, as part of the Inflation Reduction Act (IRA), Congress imposed a fee on some emissions of methane by the oil and gas industry. The IRA was most notable for providing $369 billion in tax credits and direct spending to support clean energy. While the IRA was a dramatic step forward in climate policy, Congress has yet to speak regarding EPA's authority to regulatory emissions of CO_2, the most important of the greenhouse gases.

As this book goes to press, the Biden Administration is in the midst of developing new rules to limit emissions under the Clean Air Act. Some of the initiatives are relatively straightforward, such as those imposing tighter limits on carbon emissions from vehicles and on methane emissions from the oil and gas industry. The most fraught area involves emissions from power plants, which were the subject of a recent Supreme Court opinion.

I. EPA's Regulatory Authority

The Supreme Court confronted the issue of EPA's regulatory authority in *Massachusetts v. EPA*,[1] a decision of historic importance in which the Supreme Court first encountered the issue of climate change. Although George W. Bush had endorsed limitations on carbon emissions in the 2008 campaign, he reversed course soon after taking office. During his two terms as President, the federal government obdurately resisted taking action on climate change.

A group of state and local governments, joined by thirteen leading environmental organizations, sought to force the Bush Administration's hand in regulating carbon. They invoked a provision in the Administrative Procedure Act that requires agencies to give "any interested person the right to petition for the issuance, amendment, or repeal of a rule."[2] Specifically, they petitioned EPA to regulate greenhouse gases under section 202(a)(1) of the Clean Air Act, which applies to air pollution from new vehicles.[3] This petition set the stage for what turned out to be a crucial Supreme Court ruling on greenhouse gas regulation.

---

[1] 549 U.S. 497 (2007).

[2] Administrative Procedure Act (APA) § 4(e), 5 U.S.C. § 553(e).

[3] 42 U.S.C. § 7521(a)(1).

Section 202(a)(1) requires the Administrator of EPA to issue emissions standards for new motor vehicles regarding certain air pollutants. The test for coverage is whether air pollutants "in his judgment cause, or contribute to, air pollution which may reasonably be anticipated to endanger public health or welfare."[4] The statute contains a very similar provision governing emissions from stationary sources such as power plants and factories. A key issue, then, was whether the term "air pollutant" included greenhouse gases.

It is worth taking a moment to unpack section 202(a)(1). The trigger for regulation (the so-called "endangerment finding") is highly precautionary. Rather than requiring that EPA *prove* that pollution is causing actual current harm, it only has to prove that a pollutant "may reasonably be anticipated to endanger" health or welfare. One crucial issue in the litigation was the extent of EPA's discretion to refrain from beginning the regulatory process at all. Is EPA free to forego regulation simply by declining to form any judgment about the dangers posed by a particular pollutant? Section 202 is unclear about this. The statute says the Administrator of EPA "shall" issue (not "may issue") standards for pollutants, indicating a lack of discretion. But regulation is required only if "in his judgment" the air pollutant poses a risk, which arguably introduces an element of discretion by requiring regulation only if EPA actually formulates a judgment on the matter. Instead, perhaps it can simply turn aside and decline to make a judgment one way or the other.

After considering the matter for almost four years, EPA denied the rulemaking petition on two grounds. First, it contended that it lacked regulatory authority over greenhouse gases because they are not "air pollutants" within the meaning of the statute. This was contrary to the position EPA had taken under the Clinton Administration, although it had not actually gotten to the point of making an endangerment finding. Administrative agencies are, however, entitled to change their minds. Second, EPA stated that even if it did have authority to regulate greenhouse gases, it would exercise its discretion not to exercise that authority. EPA gave two reasons for declining to exercise whatever regulatory authority it might have: (a) because of residual uncertainty over whether these gases cause global climate change, and (b) because other approaches to addressing climate change, such as international negotiation, were preferable.

Much of the Court's opinion is devoted to the issue of whether any of the petitioners had standing to challenge EPA's decision. If none of them did, the case would simply be dismissed for lack of

---

[4] *Id.*

jurisdiction without any ruling on the merits. The standing issue in the case is complicated, and we will analyze it in later in this chapter. For the present, it is enough to say that the Court found that at least one of the plaintiffs did have standing. Consequently, the Court was able to proceed to the merits of the case. Having found that it did have jurisdiction over the case, the Court then concluded that the statute plainly covered greenhouse gases. The statute contains a definition of "air pollutant" that includes any substance introduced into the air, language that certainly includes greenhouse gases. The court remarked that "[o]n its face, the definition embraces all airborne compounds of whatever stripe, and underscores that intent through the repeated use of the word 'any.' " Moreover, the Court said, "[w]hile the Congress that drafted § 202(a)(2) might not have appreciated the possibility that burning fossil fuels could lead to global warming, they did understand without regulatory flexibility, changing circumstances and scientific developments would soon render the Clean Air Act obsolete." In the Court's view, the "broad language of § 202(a)(1) reflects an intentional effort to confer the flexibility necessary to forestall such obsolescence." Given the strongly pro-environmental tenor of those times, the Court was surely right that the 1970 Congress would have endorsed using the statute to address climate change.

Because it found that EPA did have authority to regulate greenhouse gases if they endangered human health or welfare, the Court then turned to the question of the agency's discretion not to exercise that authority. Agencies have considerable discretion over whether to exercise their powers, since they must set priorities, so this was probably EPA's best argument against regulating. Nevertheless, the Court disposed of this argument in short order. According to the Court, "[w]hile the statute does condition the exercise of EPA's authority on its formation of a 'judgment', that judgment must relate to whether an air pollutant "cause[s], or contribute[s] to, air pollution which may reasonably be anticipated to endanger public health or welfare." Or, "[p]ut another way, the use of the word 'judgment' is not a roving license to ignore the statutory text," but merely "a direction to exercise discretion within defined statutory limits." Hence, EPA could avoid engaging in regulation only if it found that "greenhouse gases do not contribute to climate change or it provides some reasonable explanation as to why it cannot or will not exercise its discretion to determine whether they do."

Turning to EPA's explanation of its reasons for declining to regulate, the Court said they had nothing to do with whether greenhouse gases cause climate change, and "[s]till less do they amount to a principled reason for declining to form a scientific judgment." Nor was the existence of residual uncertainty a valid justification unless the uncertainty was so great that it prevented EPA from making a reasoned judgment. After all, the Court might have added, the statute itself plainly presumes that some uncertainty might exist when it keys regulation to whether a pollutant "may be reasonably anticipated" to endanger public health.

In dissent, Justice Scalia argued that the Court had mistaken the issue before it. He agreed that, if EPA makes a judgment under this section, the judgment must relate purely to public health risks. "But," he said, "the statute says *nothing at all* about the reasons for which the Administrator may *defer* making a judgment—the permissible reasons for deciding not to grapple with the issue at the present time." Indeed, he said, "[t]he reasons EPA gave are surely considerations executive agencies *regularly* take into account (and *ought* to take into account) when deciding whether to consider entering a new field: the impact such entry would have on other Executive Branch programs and on foreign policy."

Justice Scalia had a point, given the normally broad discretion of agencies to set their own priorities. But the majority might well have been skeptical of EPA's purported justifications. As to the question of uncertainty, the government itself had cited with approval a report of the National Academy of Science that revealed that the evidence for anthropogenic climate change was strong. And, as to the idea that international negotiations were a better approach, the Bush Administration had shown no inclination whatsoever to negotiate for international emissions limitations. The reality was that the Bush Administration did not want to use the Clean Air Act or any other mechanism to regulate because it believed that expanded production of fossil fuels was more important than preventing climate change. That policy decision was simply not consistent with the priorities of the Clean Air Act in favor of precautionary responses addressing risks to the public health and welfare.

The ultimate question in *Massachusetts v. EPA* was whether Congress intended to allow EPA to ignore potentially harmful air pollutants for policy reasons not discussed in the statute. Once it actually began a rulemaking proceeding on the subject, the statute made it clear that EPA would not be allowed to consider those policy reasons in deciding whether or not to regulate. The statutory language and the case law both made it clear that, once EPA determines public health or welfare are endangered, it must issue a vehicle emission standard. One might well ask what difference it should make that EPA had not yet begun the rulemaking proceeding. The only distinction between the situation before beginning a rulemaking and the decision after a rulemaking is that the cost of the proceeding has not yet been incurred, so EPA had to decide if this would be a wise use of its resources. EPA did not, however, phrase its refusal to form a judgment about the risks of greenhouse gases in terms of other resource priorities. Indeed, given the exceptional strength of the scientific consensus about the dangers of climate change, declining to form a judgment about the realities of climate change would seem less like an exercise of discretion by EPA than a case of willful blindness.

It is important to note that the Court did not entirely strip EPA of discretion. According to the Court, it "no doubt has significant latitude as to the manner, timing, content, and coordination of its regulation with those of other agencies." Moreover, EPA could "avoid taking further action" by providing "some reasonable explanation as to why it cannot or will not exercise its discretion" to form a judgment about climate change. In particular, it might determine that "the scientific uncertainty is so profound that it precludes EPA from making a reasoned judgment as to whether greenhouse gases contribute to global warming." Still, it seems clear that any remaining discretion must relate to the problem of forming a judgment, not to extrinsic reasons why EPA might prefer not to regulate a group of dangerous air pollutants.

By ruling that EPA did have regulatory authority regarding greenhouse gases and that its decision on whether to regulate these pollutants could only be based on scientific evidence, the Supreme Court's ruling set EPA on the path toward establishing federal climate policy. In the aftermath of the decision, EPA actually proposed an endangerment finding that was never made public, but these initial efforts were blocked by the White House. The government continued to stall throughout the remaining years of the Bush Administration. After President Obama came into office, however, it was clear that EPA would soon be exercising its newfound authority.

II. Regulation Under the Clean Air Act

Once EPA was directed to base its decision purely on science, there was little doubt about how it would ultimately rule. As we saw in Chapter 1, the scientific evidence on the link between greenhouse gases and climate change is compelling, as is the evidence about the risks involved in raising greenhouse gas levels in the future. Nevertheless, EPA faced some considerable challenges. First, it had to document the science in sufficient detail to stand up to attacks from industry and conservative state governments in court. Second, once it had decided to regulate greenhouse gases, it had to figure out how to do so within the confines of the Clean Air Act. This section tells the story of how EPA has tried to cope with these difficulties and how the courts have reacted. The first initiatives were taken under President Obama, then President Trump rolled them back, and President Biden is now attempting to set a new course.

A. The Endangerment Finding

This first step toward regulation was a finding of endangerment under section 202(a)(1). On remand, to no one's surprise, EPA made a formal finding that greenhouse gas emissions endanger human health or welfare. Under the Administrative Procedure Act, a court can set aside such a finding only if it is arbitrary or capricious. In considering such an issue, the court does not make its own judgment about the evidence, something well beyond its expertise. Instead, it probes the decision-making record to determine whether the agency gave a reasoned explanation of its judgment based on the evidence in the record. Challengers will attempt to poke holes in the agency's logic or identify evidence that was ignored by the agency.

These challenges came before the D.C. Circuit in *Coalition for Responsible Regulation, Inc. v. EPA.*[5] The challengers raised several issues about the EPA finding. First, they argued that EPA, in effect, had delegated its judgment to other bodies such as the IPCC and the National Research Council by relying on their scientific assessments. Clearly, the statute requires EPA to form its own judgment rather than blindly adopting the views of some other body. But EPA cited a large volume of evidence, not just the ultimate conclusions of these expert bodies, so this argument was something of a stretch. Indeed, the court rejected the argument as "little more than a semantic trick." In reality, the court said, EPA had merely made normal use of the existing scientific literature, and carefully evaluated the quality of these sources before relying on them.

Second, the challengers argued that the scientific evidence in the record did not support the finding of endangerment. The court carefully recounted the basis for this finding in the scientific evidence, concluding that there was substantial evidence that climate change endangers health and welfare. Industry argued, however, that there was too much uncertainty to support EPA's conclusion. In rejecting the industry argument, the court stressed that the statute is precautionary in nature and that to wait for certainty would block preventive regulation. In the court's view, the statute "requires a precautionary, forward-looking scientific judgment," so as "to prevent reasonably anticipated endangerment from maturing into concrete harm." It is worth noting that this

[5] Coalition for Responsible Regulation, Inc. v. EPA, 684 F.3d 102 (D.C. Cir. 2012). The Supreme Court granted cert. on another issue in the case and reversed in part on that issue in Utility Air Regulatory Group v. EPA, 573 U.S. 302 (2014).

approach resonates with the Precautionary Principle found in international environmental law, though the court did not say so.

Two other arguments that were made before the court are also worth mentioning. First, some of the parties commenting on the rule argued that EPA had improperly based its endangerment finding on possible harm to foreign populations. As we saw in Chapter 1, this is a concern that has also been raised concerning the social cost of carbon. EPA made it clear, however, that although it considered the global effects of greenhouse gases, it did so only in the course of determining potential domestic harm. EPA thus dodged the issue of whether it could have relied on harm to human health or welfare outside the United States if it had wanted to do so. A second question concerned the relevant time period for assessing danger. Time scale is clearly a critical issue in assessing the impacts of climate change. Rejecting arguments for focusing only on current impacts, EPA's analysis was based on the next few decades, in some cases extending up to the end of the century.

Note that, because it was making a qualitative assessment of the degree of danger, EPA was not required to quantify the probability or to set a discount rate, two major issues in the economic analysis of climate change. The challengers did argue that EPA had failed to quantify the point at which greenhouse gases pose a danger to public health, the specific types of harms, or the risks and impacts of climate change. But the court rejected the idea that EPA was required to set a numerical threshold in finding endangerment. "Quite the opposite," the court said, the endangerment finding requires a "case-by-case, sliding-scale approach to endangerment."

As discussed later in this chapter, the Supreme Court did grant review over (and partially reverse) another portion of the D.C. Circuit's judgment, but it refused to review the endangerment finding. The Trump Administration was under some pressure from conservatives to reopen the endangerment finding, but it never did so, apparently because of concerns about the litigation risks involved. Thus, the endangerment finding stands intact.

Once it had decided to make a finding of endangerment, EPA was then faced with the question of how to go about regulating greenhouse gases. This was a relatively straightforward issue in terms of vehicle emissions. Section 202 required EPA to impose standards for emissions from new motor vehicles once it had found endangerment, and EPA proceeded to do so without any huge difficulty. As discussed in the chapter on state regulation, the car industry was already under pressure because of regulations adopted in California, so EPA was not writing on a blank slate.

But it was more difficult to know how to approach emissions from stationary sources like power plants and factories. The main mechanism that is normally used for regulating emissions from stationary sources begins with establishment of national air quality standards by EPA after making an endangerment finding under section 108(a)(1).[6] (Note that this is a different section and a distinct endangerment finding from the one EPA had already made for vehicle emissions under section 202(a)(1).) States then submit implementation plans that detail how they will meet the air quality standard. But this mechanism is at best a very awkward fit for greenhouse gases. Given the global nature of greenhouse gases, it would be impossible for any state to control greenhouse gas concentrations within its own territory. Thus, it cannot possibly devise a plan that would enable it to meet a national standard for greenhouse gas concentrations. Consequently, EPA looked elsewhere in the statute for provisions that it could use to control greenhouse gases.

EPA took a long, careful look at the statute and decided to base its strategy on some provisions that do not require setting national air quality standards. These provisions were triggered by the finding of endangerment under section 202(a)(1), without the need for a separate endangerment finding under section 108 that would have triggered establishment of national air quality standards. If the distinctions between the two endangerment findings seems like an arcane technicality, it was only the start of the technical issues that EPA encountered in applying the Clean Air Act. Even those provisions that EPA did use were not always easy to apply. They had been written with conventional air pollutants in mind and adapting them to deal with CO_2 raised problems. At times, as we will see, some of those problems were quite technical, relating to numerical thresholds in the statutes or to peculiarities of drafting history. Nevertheless, EPA did the best it could with the statutory material it had to work with, with mixed results in the Supreme Court.

B. New Sources

EPA utilized two statutory provisions to deal with emissions from new stationary sources. In general, environmental statutes are prone to give the federal government more control over standards for new sources than for existing ones. Congress wanted to avoid a race to the bottom whereby states would compete for new industry by lowering standards.

The first EPA effort to extend the greenhouse gas regulatory regime to stationary sources came at the same time as the

---

[6] 42 U.S.C. § 108(a)(1).

endangerment finding under section 202(a)(1). This regulation utilized a portion of the Clean Air Act that was initially designed for quite another purpose: maintaining air quality in areas that were relatively pristine rather than letting it decline to the level of the national air quality standards. As it turned out, however, all areas of the country were eventually found to be cleaner than the national standards for at least one pollutant. Thus, the provisions now apply nationally. Also, given that EPA had no intention of issuing national air quality standards for greenhouse gases, these "prevention of significant deterioration" (PSD) requirements had the advantage that they were not necessarily limited to conventional pollutants for which air quality standards existed. Instead, EPA believed, the PSD requirements came into play once emissions of greenhouse gases from new vehicles were subject to regulation.

The key operative provision is section 165(a) of the Clean Air Act,[7] which applies to any "major emitting facility" constructed in an area subject to PSD protection (which turns out to include the entire country). Under section 165(a)(4), each such major emitter must use the "best available control technology [BACT] for each pollutant subject to regulation under this chapter emitted from, or which results from, such facility." In applying this provision, EPA was faced with two major coverage issues. First, what facilities are covered by the statute? Second, once a facility is covered for whatever reason, are greenhouse gases among the pollutants for which BACT is required?

What facilities are covered is controlled by another provision, section 169(1),[8] which defines the "major sources" that must comply with BACT. Section 169 defines a source as major if it emits or has the potential to emit a specified quantity of "any pollutant" in the amount of 100 tons per year if a facility belongs to particular set of defined industries and otherwise 250 tons per year. The first question confronted by EPA was whether coverage is triggered if a source emits more than these amounts of greenhouse gases (but not of any other pollutants). EPA answered this "trigger" issue in the affirmative, on the ground that "any pollutant" must include greenhouse gases, since they are pollutants within the meaning of the statute under *Massachusetts v. EPA*. But this left EPA with a problem. While 250 tons per year is quite a lot of most conventional pollutants, there are a huge number of relatively small sources that emit that much CO_2. EPA believed that it would be completely impractical to apply the BACT requirement to all of those facilities. It therefore adopted a "tailoring rule," which limited coverage of

---

7 42 U.S.C. § 7475(a).

8 42 U.S.C. § 7479(1).

BACT to facilities that emitted much larger amounts of greenhouse gases (at least 75,000 tons per year). Because it appeared to rewrite the statute (striking out the number 250 and replacing it with 75,000), this was an aggressive legal move. EPA defended its action based on the principle that statutes should not be interpreted to require absurd or impractical results.

This left the second question: whether facilities covered because of their emissions of other pollutants (so-called "anyway" sources) had to use BACT for greenhouse gases. An example would be a factory that emitted more than 250 tons of nitrogen oxides per year. Again, the agency answered yes. Here the statutory language was clearer. Section 165(a)(4) refers to any pollutant regulated under "this chapter," and chapter 85 of 42 U.S.C. contains the entire Clean Air Act. Thus, greenhouse gases from "anyway" sources had to be subject to BACT, because the Supreme Court had already held that they were subject to regulation under subchapter II of chapter 85, which deals with vehicles.

In a 5–4 decision, the Supreme Court reversed EPA's triggering and tailoring rules in *Utility Air Regulatory Group v. EPA*.[9] In an opinion by Justice Scalia, the Court concluded that the agency should have realized its broad definition of major facilities was completely untenable. The term "air pollutant" would normally encompass greenhouse gases, per *Massachusetts v. EPA*, but in this case that interpretation would make no sense. Even EPA agreed that interpreting the statute to cover thousands of additional sources would be absurd. EPA "lacked authority to 'tailor' the Act's unambiguous numerical thresholds to accommodate its greenhouse-gas-inclusive interpretation of the permitting triggers." "Instead," Justice Scalia continued, "the need to rewrite clear provisions of the statute should have alerted EPA that it had taken a wrong 'interpretative' turn." Given that EPA's numerical revision was invalid, its interpretation of the trigger requirement would mean coverage for "millions of small sources—including retail stores, offices, apartment buildings, shopping centers, schools, and churches." The Court rejected such an "enormous and transformative expansion in EPA's regulatory authority without clear congressional authorization."

In terms of "anyway" sources, however, the Court concluded that EPA was correct: once a source is classified as "major" because of its emission of conventional pollutants, it must use BACT for greenhouse gases. Given the specificity of the statutory language in covering all pollutants regulated anywhere in the Clean Air Act, it is hard to see how the Court could have ruled otherwise.

---

[9] 573 U.S. 302 (2014).

Justice Breyer, joined by Justices Ginsburg, Sotomayor, and Kagan, dissented. They found it peculiar that the Court was unwilling to allow EPA to modify the statutory definition of major source whereas the Court required it to modify the statutory definition of air pollutant. Once the Court admitted that the statute could not be applied literally, why was its rewrite of the statute to say "any pollutant (except greenhouse gases)" better than the agency's rewrite to "250 tons (except 75,000 tons for greenhouse gases)"? That argument, however, cogent, carried no weight with the Court. Perhaps inserting qualifying language that limited coverage seemed less like rewriting the law than substituting a different numerical limit.

In another dissent, Justices Alito and Thomas argued that *Massachusetts v. EPA* had erred in holding that the Clean Air Act covers greenhouse gases. For that reason, they would have struck down the "anyway" rule as well. None of the other Justices joined this opinion, and they seem to have given up on the idea of overruling. *Massachusetts v. EPA* subsequently.

The *UARG* decision did not have a dramatic impact on the effectiveness of the PSD rules, because at least 85 percent of GHGs come from "anyway" facilities. The Trump Administration made no effort to repeal PSD coverage for "anyway" plants. Perhaps the reason was the fact that the rule has already been upheld by the courts, which makes it more difficult to make a cogent case for repeal. Or perhaps the reason was an assumption that, in the absence of pressure from EPA, states may not find it hard to issue fairly toothless permits.

In terms of its broader implications, the ruling did raise some serious concern, however, about how open the Court will be to EPA's efforts to adapt other portions of the Clean Air Act to greenhouse gases. But it may be a mistake to read too much into the majority's rejection of the tailoring rule. By appearing to modify the specific numerical limits in the statute, the tailoring rule was an unusually aggressive legal move by EPA, which even some sympathetic observers thought was very risky.

After the PSD regulations issued above, in 2015 the Obama EPA issued standards covering new electric power generators under section 111 of the statute.[10] Section 111(b) directs the Administrator to promulgate "standards of performance" governing emissions of air pollutants by "new" stationary sources, i.e., sources constructed or modified after the effective date of pertinent regulations. The term "modification" means any physical change or change in the method of operation which significantly increases the net amount of any air

---

[10] 42 U.S.C. § 7411.

pollutant. The emission standard must reflect "the degree of emission reduction achievable through the application of the best system of continuous emission reduction which . . . the Administrator determines has been adequately demonstrated." The key terms to focus on here are "best system of continuous emission reduction" (BSER) and "adequately demonstrated."

EPA set a standard of emission limit of one ton of CO_2 per megawatt-hour for natural gas plants providing baseload power (that is, running outside of peak power demand). EPA found that the best available technology for new coal-fired plants consists of a supercritical pulverized coal unit using carbon capture and storage (CCS) to eliminate about twenty percent of carbon emission. The final standard for coal plants based on this technology was an emission limit of 2.8 tons of CO_2 per megawatt-hour. There will undoubtedly be strong disputes over whether carbon capture has been "adequately demonstrated."

Because no new coal fired plants are currently in the works, EPA suggested that this requirement would have very little cost, at least for the next decade. The fact that no new plants are currently in the works means that there is time for industry to improve CCS technology, a relevant factor in determining whether that technology has been "adequately demonstrated" at present. This rule—sometimes referred to as the New Source Rule—has been largely untested in litigation[11], and was not part of the challenge in *West Virginia v. EPA*.[12] Moreover, despite efforts during the Trump Administration to modify the 111(b) rules, these efforts were rejected by the D.C. Circuity in 2021. As a result, the 2015 Obama Administration's rule remains in place.

C. Existing Sources

This brings us to the most hotly contested issue about EPA's climate change initiatives: its power to cut emissions from existing power plants that use fossil fuels. In order to regulate existing power plants—especially existing coal-fired plants—EPA turned to section 111(d) of the Clean Air Act,[13] a previously obscure provision. Section 111(d) provides that, with an exception we will discuss below, EPA can require states to submit plans to control emissions from existing plants once it has issued a standard for new sources in the same category under section 111(b). As we saw in the previous section, EPA did issue new source standards for greenhouse gases from power

---

[11] *See* Las Brisas Energy Ctr. v. EPA, 2012 U.S. App. Lexis 25535 (D.C. Cir. Dec. 13, 2012).

[12] 142 S. Ct. 2587 (2022). The *West Virginia* case involved only a related rule applying to existing sources.

[13] 42 U.S.C. § 7411(d).

plants under section 111(b), so section 111(d) came into play. If a state fails to submit a plan, EPA must submit its own enforceable plans for that state. The plans are supposed to be based on the standard of performance for the industry—that is, the best "system of continuous emission reduction" (BSER) that has been "adequately demonstrated" in terms of existing plants in that state. A crucial issue involved the scope of the term "system"—does it include only plant specific emission limitations measures, or could a system be defined more broadly?

The Obama Administration's section 111(d) regulation was known as the Clean Power Plan. In defining BSER the Clean Power Plan took a different approach for existing plants than EPA had used for new plants, given that existing coal plants could not be retrofitted to use an entirely different technology (supercritical pulverized coal plus partial carbon capture). It determined that the best system of emission reduction for existing units consisted instead of three building blocks: (1) efficiency improvements in coal-fired plants, (2) substitution of natural gas generation for coal-fired generation when feasible, and (3) increased use of renewables. Because the power system is organized around three interconnected grids (East, West, and Texas), EPA determined what emissions reductions could be feasibly achieved nationally by applying each building block in each of the three grid areas. It then used the least common denominator for each building block to set a national emissions reduction standard. Finally, EPA applied the building blocks to each state depending on their own mix of power sources—for instance, states that already made high use of natural gas and little use of coal obviously would find it more difficult to achieve reductions by further switching away from coal.

Under section 111(d), it is up to states to issue plans for meeting the target, with an EPA-issued plan as a backup if they fail to act. States were required to achieve interim targets between 2022 and 2029, and then the final target in 2030. EPA seemed to be anxious to give states as much flexibility as possible. The regulation provided a bewildering set of options for states. For instance, states could either set targets in terms of tons per megawatt or total carbon emissions. They could reach the emissions targets by setting emissions limits for the fossil fuel generators, leaving it to the generators to figure out how to meet the standard, or they could adopt a regulatory scheme embodying the three building blocks. They were encouraged but not required to use emissions trading to attain the standards more cheaply.

EPA estimated that the Clean Power Plan had public health and climate benefits worth an estimated $34 billion to $54 billion per year in 2030, far outweighing the costs of $8.4 billion per year. The public health co-benefits were a major part of the equation, because reducing the use of coal would eliminate a lot of harmful particulates from the air. The plan was expected to cut carbon emissions from power plants by about a third.

Even before EPA had formally issued the regulations, lawsuits were filed to try to halt the process. Unusually, the Supreme Court granted a stay even though the D.C. Circuit had not yet ruled on the merits. There were two fundamental challenges to EPA's authority, one of which ultimately reached the Supreme Court.

The first fundamental challenge relates to the definition of the BSER. Typically, new source regulations under section 111 have been based on plant-specific systems of pollution control such as scrubbers. With respect to building blocks 2 and 3, EPA had to contend with arguments that the "system" of control cannot include "beyond the fence line" measures such as increases in use of natural gas generation and renewables. EPA's proposed rule originally included energy efficiency as a fourth building block. Unlike the other building blocks which were supply-side regulation, the efficiency block was a demand-side effort to reduce electricity consumption; it therefore extended the regulation's purview beyond electricity suppliers to their customers. That building block was abandoned largely because of fears that it would not qualify as part of a "system" of control. Defining the "system" of pollution control to encompass changes in the amount of electricity introduced into the grid could have been a departure for EPA, which normally defines it as a type of pollution control equipment at the specific emitting facility. In effect, the Clean Power Plan treated all the power generators on the state grid as part of a single unified source. This makes a certain amount of sense because of the way the grid operates—it has been called the world's most complicated machine—and because of the practicalities of controlling carbon. But it proved too innovative for courts to accept.

The other challenge was more arcane. Section 111(d) was initially designed as a gap filler for situations that were not covered by section 108 (pollutants covered by national air quality standards) or section 112 (toxic chemicals).[14] In 1990, as part of an overhaul of the statute, Congress amended 111(d) to clarify the "section 112 exclusion" for toxic chemicals. But, unfortunately, the two houses of Congress passed different versions. Normally, when this happens a Conference Committee irons out the disagreement by picking one version or the other, or by drafting a compromise. But something went wrong in this case, and both versions somehow found their ways into the law. The Senate version calls for regulation of any air

[14] 42 U.S.C. § 7412.

pollutant that is not covered by section 108 or listed under section 112. Since carbon dioxide falls in neither category, it is clearly covered by section 111(d). But the House language is somewhat different. It excludes any pollutant "emitted from a source category" that is regulated under section 112. Thus, read literally, the House version means that if arsenic emissions from some industry are regulated under section 112, that industry is immune from regulation under section 111(d) for all other pollutants. The difference in language between the House and Senate versions created an interpretative conundrum. Although industry had hopes of leveraging this drafting glitch to torpedo any use of section 111(d), the D.C. Circuit later rejected the argument and the Supreme Court declined to review that holding.

After oral argument over the Clean Power Plan, it appeared that the D.C. Circuit was likely to rule in EPA's favor on these key legal issues. But the 2016 presidential election intervened before the court could rule. President Trump issued an executive order directing EPA to reconsider the rule, and his accompanying public remarks suggested strongly that the result would be to repeal the rule (not that this was a huge surprise). The government then petitioned the court to hold the case in abeyance while it began a new rulemaking. On April 28, 2017, the D.C. Circuit granted the Trump administration's request to suspend lawsuits against the Clean Power Plan rule. Subsequently, on October 16, 2017, the EPA proposed repealing the CPP and then, on August 21, 2018, EPA issued its proposed replacement, the Affordable Clean Energy (ACE) rule.[15] The CPP was finally repealed and replaced with the ACE on June 19, 2019.[16]

Unlike the Clean Power Plan, which was designed to ratchet down emissions from power plants by 32% below 2005 levels by 2030,[17] the ACE did not establish an absolute greenhouse gas emissions reduction target. Rather, the EPA suggested that "along with additional expected emissions reductions based on long-term industry trends" the rule *could* result in emissions reductions "as much as 35% below 2005 levels" by the year 2030.[18] ACE operated by

---

15 *See* EPA, *Fact Sheet: The Affordable Clean Energy Rule*, https://perma.cc/NN9V-Q2WZ.

16 *EPA Finalizes Affordable Clean Energy Rule, Ensuring Reliable, Diversified Energy Resources While Protecting Our Environment*, EPA (June 19, 2019), https://www.epa.gov/newsreleases/epa-finalizes-affordable-clean-energy-rule-ensuring-reliable-diversified-energy [https://perma.cc/A3Z3-UYPE] (suggesting that ACE will reduce CO_2 emissions by 11-million short tons—less than 1% of current U.S. emissions, SO_2 emissions by 5700 tons, NOx emissions by 7100 tons, PM2.5 emissions by 400 tons, and mercury emissions by fifty-nine pounds).

17 *See Fact Sheet: Overview of the Clean Power Plan*, EPA, https://perma.cc/VBE2-9ML7.

18 *Id.*

mandating heat-rate efficiency improvements ("HRIs") at individual facilities; the rule did not mandate net emissions reductions by state or even by facility.

The Trump EPA contended that the statute unambiguously barred any regulation beyond efficiency improvements at individual coal-fired power plants. The D.C. Circuit then rejected the argument that the statute was unambiguous, vacated the Trump rule, and left it to EPA under President Biden to formulate a new regulation. Again surprising observers, the Supreme Court agreed to hear the case even though EPA had not had time to formulate a new regulation. Although the utility industry did not seek review, states and coal mining companies argued EPA was limited to regulations that took effect at individual coal-fired power plants. One party also argued that EPA was limited to issuing general guidelines for states.

The issue reached the Supreme Court in *West Virginia v. EPA.*[19] In a 6–3 decision, the Court upheld that portion of the Trump Administration's ACE rule that rescinded the Clean Power Plan. The Court's rationale was that EPA's claim of authority to issue the plan raised a "major question." Under what is called the major questions doctrine, an agency must demonstrate a clear delegation of authority from Congress when a regulation involves a major question. What constitutes a major question has not been clearly defined, but Chief Justice Roberts's majority opinion pointed to four factors:

1. ***Stark departure from past practice and regulatory norms.*** The agency's interpretation of the statute was "not only unprecedented; it also effected a 'fundamental revision of the statute, changing it from [one sort of] scheme of . . . regulation' into an entirely different kind." Moreover, section 1111(d) was an obscure and rarely used provision. The Court considered it unlikely that Congress would conceal such a major grant of power in such an obscure place.

2. ***Breadth of the claimed authority.*** Under EPA's view of the statute, Roberts says, "Congress implicitly tasked it, and it alone, with balancing the many vital considerations of national policy implicated in deciding how Americans will get their energy." Congress needs to say so clearly if that's what it intends.

3. ***Lack of relevant expertise.*** In the Court's view, EPA lacked expertise on running the electricity system. This was shown, the Court said, by a budget request in

[19] 142 S. Ct. 2587 (2022).

which EPA said it needed additional funding in order to develop such expertise.

4. ***Congressional consideration and rejection.*** Congress considered and rejected multiple efforts to create a cap-and-trade scheme for carbon.

In the end, the Court objected to EPA having the power to establish what share of electricity should come from different sources, including the power to eliminate coal entirely if it chose. It also viewed the Clean Power Plan as akin a nationwide cap and trade system. These may not have been fair characterizations of EPA's actions. The major question doctrine itself is controversial and has been harshly criticized. Nevertheless, the Court's rejection of the Clean Power Plan is a fact that future EPA efforts must contend with.

Given the way that the Court defines the major questions doctrine, *West Virginia v. EPA* does not seem to threaten any of the other climate regulations discussed in this chapter. It does pose EPA with a key question: what regulations of power plants are possible after this ruling?

If the bad news is the Court's rejection of the Obama Administration's broad view of section 111(d)m, the good news is that it did not embrace the Trump Administration's extremely narrow view either. Unlike the Trump Administration, the Court left open the possibility that EPA could use cap and trade, at least as a compliance method. The Court also rejected the view that regulation was strictly limited to requirements "inside the fenceline." The Court said, "We have no occasion to decide whether the statutory phrase 'system of emission reduction' refers *exclusively* to measures that improve the pollution performance of individual sources, such that all other actions are ineligible to qualify as the BSER."

The Court's ruling in the case was not a surprise, given the makeup of the bench, and EPA presumably had already embarked on a search for alternative ways to implement section 111(d). One possibility that could result in significant reductions would be to require coal-fired power plants to add natural gas to their power mix. But at this writing, what path EPA will take remains unknown.

Even under an expansive reading of section 111(d) like the Obama Administration's, it has always been clear that there are limits on what EPA can expect to accomplish without new legislation from Congress. *West Virginia v. EPA* underlines that point. Despite all the barriers to congressional action, it will be indispensable at some point to achieve emission reductions of the scale needed by midcentury. The recent Inflation Reduction Act was a significant step by Congress to enter the climate arena via large-scale spending.

It is worth noting that there are other statutory provisions that potentially provide EPA with authority to regulate GHG emissions. One possibility is section 115 of the Clean Air Act.[20] Section 115 directs EPA to require states to revise their pollution control plans to prevent or eliminate U.S. emissions that cause environmental problems in other countries. This authority might provide another route for EPA to address carbon emissions. But it seems almost certain that the Supreme Court would find that use of section 115 would involve the same kinds of major questions that the Court identified in the *West Virginia* case. Recourse to other sections of the Clean Air Act as a source of authority for sweeping climate action might raise similar problems.

III. Beyond the Clean Air Act: Climate Change and Other Federal Statutes

EPA's efforts to regulate carbon emissions under the Clean Air Act constitute the most important federal response to climate change. But other federal agencies have also had to confront the issue of climate change. This section discusses some of their efforts to do so.

A. Environmental Impact Statements Under NEPA

Some brief background on the National Environmental Policy Act (NEPA) may be helpful. NEPA was the opening shot in what could be considered the "long decade" from 1969 to 1981 when Congress enacted nearly all of the modern environmental statutes. The most significant provision of NEPA is undoubtedly section 102(2)(c).[21] The primary purpose of this provision is to force agencies to take environmental factors into consideration when making significant decisions. The crucial language of section 102(2)(c) requires federal agencies to include in every "recommendation or report" on "major Federal actions significantly affecting the quality of the human environment" a detailed statement covering:

(i) the action's environmental impact,

(ii) any unavoidable adverse environmental effects of the action,

(iii) alternatives to the proposed action.

The subsection goes on to require the decision-making federal agency to consult other agencies with jurisdiction over, or special expertise

---

[20] 42 U.S.C. § 7415.

[21] 42 U.S.C. § 4332(2)(c).

concerning, the environmental problem involved. Copies of the resulting Environmental Impact Statement (EIS) must be circulated among federal, state, and local agencies, to the President, to the Council on Environmental Quality (CEQ), and to the public. The EIS also must "accompany the proposal through the existing agency review processes."

In essence, the statute requires the agency to prepare a detailed explanation of the environmental consequences of its proposed actions and to make that report available to higher-level agency officials, other agencies, and the public. The goal is to ensure that the relevant agency and the general public are fully informed of the potential environmental impacts of the project and the extent to which they can be limited. The Supreme Court has indicated that once an agency discloses the relevant information, the results of the EIS do not tie NEPA's hands in making the actual decision.

The EIS provision in NEPA makes no reference to judicial enforcement. Indeed, the drafter of the provision apparently did not have such enforcement specifically in mind. Nevertheless, soon after the statute was passed, it became clear that courts would become actively involved in enforcing it.

Regulations by CEQ establish procedures for determining whether an EIS is required. The process begins with an environmental assessment (EA), which is a brief analysis of the need for an EIS. The EA analyzes whether there are any significant environmental impacts that would require an EIS. The EA must also consider alternatives to the proposed action, as required by section 102(2)(E) of NEPA even if environmental impacts are not "significant." When an action would otherwise result in a significant environmental impact, the agency may be able to mitigate the impact in order to avoid the need for an EIS. If the agency decides not to prepare an EIS, it must make a "finding of no significant impact" (FONSI) available to the public.

The key question for our purposes is how climate change factors into NEPA. The leading case is *Center for Biological Diversity v. Nat'l Highway Traffic Safety Admin.*[22] At issue was a Bush Administration rule setting fuel efficiency (CAFE) standards for small trucks and SUVs. The court faulted the agency for failing to consider how alternatives to its preferred rule could be more effective in reducing greenhouse gases. In the court's view, the "impact of greenhouse gas emissions on climate change is precisely the kind of cumulative impacts analysis that NEPA requires agencies to conduct." Admittedly, any given rule setting fuel efficiency standards might have a minor effect by itself, but these rules are "collectively

[22] 538 F.3d 1172 (9th Cir. 2008).

significant actions taking place over a period of time." Thus, the agency "must provide the necessary contextual information about the cumulative and incremental environmental impacts of the Final Rule in light of other CAFE rulemakings and other past, present, and reasonably foreseeable future actions, regardless of what agency or person undertakes such other actions." In short, given the clear link between fuel efficiency and carbon emissions, the agency needed to include climate change in its consideration of environmental impacts.

Courts have followed this precedent in requiring consideration of climate impacts for large transportation projects and transmission lines, as well as related rulemakings. The Trump Administration was repeatedly reversed in cases relating to oil and gas production and coal mining on public lands for failure to take into account the climate effects of these activities. The Federal Energy Regulatory Commission also ran into solid judicial resistance because of its refusal to consider climate impacts of gas pipelines and gas export projects.

One of the remaining areas of uncertainty is when climate emissions rise to the level where their impacts are "significant" and require a full-scale EIS. CEQ draft guidance issued under Obama rejected the view that emissions can be ignored if the amount involved in the agency's decision is small, because by its nature climate change involves large consequences as a result of a multitude of relatively small actions. However, the extent of the discussion of climate change should be proportional to the amount of emissions, and a quantitative analysis of climate consequences is not needed if the amount of greenhouse gas emissions at issue is below 25,000 metric tons of carbon. The guidance also made the fairly obvious but important point that the agencies must also consider how climate change could amplify other environmental impacts of a project. An Executive Order by Trump directed the CEQ to rescind the guidance and issued new guidance that minimized Agencies' obligation to consider greenhouse gas emissions in the EIS process. In 2021, however, the Biden Administration rescinded the Trump-era guidance and, at the time of writing, the CEQ is in the process of revising and updating the 2016 Obama-era guidance in line with President Biden's Executive Order, *Protecting Public Health and the Environment and Restoring Science to Tackle the Climate Crisis.*[23]

Despite the efforts by the Trump Administration to minimize the extent of this obligation, environmental impact statements must address a project's carbon emissions. That remains the rule in the Ninth Circuit, which has been followed by other courts. This fits with

---

[23] Executive Order 13990, *Protecting Public Health and the Environment and Restoring Science to Tackle the Climate Crisis*, 86 FED. REG. 7037 (Jan. 25, 2021).

NEPA's general emphasis on ensuring that agencies take a hard look at the environmental consequences of their actions. It is possible, however, that some other courts might accept counterarguments that the causal link between a given project and climate impacts is too attenuated or uncertain to warrant discussion. But, as will be discussed, similar arguments were made to the Supreme Court in the context of standing and were rejected by the Court in *Massachusetts v. EPA*, making their viability dubious in the absence of further guidance from the Supreme Court.

B. The Endangered Species Act

Some of the key impacts of climate change are ecological. The evidence already shows significant effects of climate change on animal and plant life. According to climate scientists, given the relatively small degree of global warming to date, clear impacts on species are already observable. For this reason, irreversible harm could occur to ecosystems as their resilience levels are exceeded. Indeed, some ecologists are speaking of the current era as akin to the great extinction events in the geological record.[24] Yet, specific effects on particular ecosystems are harder to predict than purely physical changes because of the complexity of biological systems.

Both the potential for extinction and the difficulties of making precise predictions pose problems for the legal system. In the United States, the key mechanism for protecting biodiversity is the Endangered Species Act.[25] A short survey of the statute is in order as a prelude to discussion of its treatment of climate change.

The first step in applying the statute is the listing of an endangered or threatened species. Under section 4 of the Act, the Secretary of the Interior (in the case of land-based and freshwater species) or the Secretary of Commerce (in the case of marine species) must "determine whether any species is an endangered species or a threatened species" based on a series of factors, including "the present or threatened destruction, modification, or curtailment of its habitat or range" and "other natural or manmade factors affecting its continued existence."[26] Climate change falls under the first category as a cause of habitat modification or destruction, as well as under the catchall category as a "natural or manmade factor."

The Act defines an "endangered species" as "any species which is in danger of extinction throughout all or a significant portion of its

---

[24] *See* ELIZABETH KOLBERT, THE SIXTH EXTINCTION: AN UNNATURAL HISTORY (2014).

[25] 16 U.S.C. § 1531.

[26] 16 U.S.C. § 1533(a) (2000).

range."[27] A "threatened species," on the other hand, is "any species which is likely to become an endangered species within the foreseeable future throughout all or a significant portion of its range."[28] In the listing of a species, the Secretary may not consider economic impacts. Rather, the Secretary must make the determination "solely on the basis of the best scientific and commercial data available to him."[29]

The effects of listing are substantial. First, the species obtains stringent protection on federal lands. Section 7(a)(2) of the Act provides that "[e]ach Federal agency shall . . . insure that any action authorized, funded, or carried out by such agency . . . is not likely to jeopardize the continued existence of any endangered species or threatened species or result in the destruction or adverse modification of [critical] habitat of such species . . ."[30] Second, section 9 of the ESA, arguably the most controversial aspect of the statute, establishes a broad prohibition against "taking" endangered species. Unlike section 7, which applies only to federal agencies, section 9 applies to "any person subject to the jurisdiction of the United States." The Supreme Court has held that the "taking" prohibition applies not only to direct killing but also to habitat modifications that proximately cause the death of members of the species.[31] Because this provision limits habitat destruction on private lands, it has been the target of attack by development interests as well as by property rights advocates.

The emblematic victim of climate change is the polar bear, whose listing as a threatened species was upheld by the D.C. Circuit in the appropriately named *In re Polar Bear Litigation*[32] case. The *Polar Bear* case involved several important issues, including potential uncertainties regarding the impact of climate change on the bears.

The litigation is a good example of the kinds of issues that are likely to be raised in cases involving the impacts of climate change. EPA had relied in part on two models of polar bear populations developed by the United States Geological Service (USGS). These models were a far cry from the sophistication of the climate models discussed in Chapter 1. One model was simply based on the past

---

27 16 U.S.C. § 1532(6) (2000).

28 16 U.S.C. § 1532(20).

29 16 U.S.C. § 1533(b)(1)(A). *See* New Mexico Cattle Growers Ass'n v. United States Fish & Wildlife Serv., 248 F.3d 1277 (10th Cir. 2001) (holding that Service may not consider economic impacts in listing decision).

30 16 U.S.C. § 1536(a)(2).

31 Babbitt v. Sweet Home Chapter of Communities For a Great Oregon, 515 U.S. 687 (1995).

32 In re Polar Bear Endangered Species Act Listing and Section 4(d) Rule Litigation—MDL No. 1993, 709 F.3d 1 (2013).

statistical relationship between the area of sea ice and polar bear populations; the other included other potential stressors as well as indications of the availability (not just the area) of the ice. The Fish and Wildlife Service itself indicated doubts about these models, since the first one relied on a dubious assumption of constant population densities while the second was in an early stage of development. The court dismissed this challenge to the regulation, however, because the agency used the models only for the limited purpose of confirming trends indicated by other evidence.

The challengers then criticized the agency for relying on the IPCC definition of "likely" (67 to 90 percent) in applying the statutory standard of whether polar bears were likely to become endangered in the foreseeable future. But the court found this claim to be based on an implausible interpretation of the agency's language when read in context. Instead, the court concluded that the agency followed the IPCC practice only when stating its confidence in climate forecasts and otherwise applied the term in a commonsensical (though not precisely defined) way.

Finally, the challengers contested the time period considered by the agency. To classify a species as threatened, the agency has to find that it is likely to become endangered in the foreseeable future. But what does the term foreseeable future mean? The agency defined "foreseeable" as forty-five years, which meant it stretched out to 2050. The agency chose this timeframe as being the period during which it could make a reliable assessment of the effect of threats on the species because climate models were in essential agreement until about that time. Presumably many more species are likely to become threatened by the end of the century if climate sensitivity or emissions levels turn out to be high.

Predictions about the future of a species already are hampered by the unpredictability of ecological systems, and uncertainty about the severity of climate change necessarily complicates the problem further. Nevertheless, the government seems to have made a thoughtful decision despite these uncertainties, helped along by a precautionary statutory standard.

Although the polar bear is a particularly obvious victim of climate change, climate change has been a factor in other listing-related decisions. It seems clear that agencies simply cannot ignore climate change in listing decisions. For instance, in *Greater Yellowstone Coalition v. Servheen*, the court struck down a decision to remove the Yellowstone grizzly bear from the endangered list for this reason. Climate change has also been a key consideration in a number of listing decisions involving marine species.[33] In addition,

---

[33] Eric Seney et al., *Climate Change, Marine Environments, and the U.S. Endangered Species Act*, 27 CONSERVATION BIOLOGY 1138, 1142 (2013).

climate change has also been an increasingly prevalent consideration in planning for species to recover from crisis levels,[34] including two species of coral.[35] The government now routinely considers the effects of climate change on a species, and climate considerations have proved particularly relevant to a number of aquatic life forms.

While there is an argument that the ESA might prohibit carbon emissions as a "taking" of endangered species, the government has rejected this approach so far. Indeed, it announced that view in a rule accompanying the designation of the polar bear as threatened. The rule was reconsidered but then reaffirmed after the Obama Administration took office. If the government had not taken this position, all or some proportion of carbon emissions might have instantly become illegal. The government is also reluctant (though it has not taken a formal public position on this) to consider carbon emissions controlled by the government as a source of "jeopardy" to endangered species. Because classifying emissions as a source of jeopardy would affect only activities conducted, licensed, or funded by the federal government, it would have less impact, but it might well eliminate drilling and coal mining on federal land.

The main argument against using the Endangered Species Act to address carbon emissions is that the statute is poorly designed to do so. The ESA is designed to deal with discrete projects or individual actions that threaten a species, not with widely dispersed human activity such as carbon emissions generated by the world's energy system. Thus, the ESA is arguably a poor fit for climate mitigation efforts. For instance, the requirement that an action *proximately* cause the death of members of the species might exclude carbon emissions. This does not mean that the Endangered Species Act has no role to play in dealing with climate change. Where species are not already irreversibly doomed by climate change, the statute could provide impetus to protect remaining populations from other threats, and to help species in overcoming barriers to migration to safer haven.

Even the Trump Administration did not attempt to dispute that climate change can jeopardize the survival of some species. Given that the courts have already upheld the application of the statute on this basis, a shift in position might have required a particularly cogent argument, since the existence of jeopardy is supposed to be a biological determination, not a policy decision. Clearly, it would not have been impossible for Trump to take that position, but it may have

---

[34] *Id.* at 1143.

[35] ROBIN KUNDIS CRAIG, COMPARATIVE OCEAN GOVERNANCE: PLACE-BASED PROTECTION IN AN ERA OF CLIMATE CHANGE 11 (2012).

been easier simply to dispute that climate change would impact on any particular species that the agency was asked to list.

IV. Judicial Review and Climate Policy

A key issue regarding all of these federal statutes is judicial review. It is relatively easy for firms in a regulated industry to challenge regulations for being too strict. But to ensure vigorous implementation of environmental statutes, it is also important for environmental groups and concerned citizens to have access to the courts. This section will discuss the doctrine of standing, which is the main potential obstacle to bringing climate change litigation.

For a party seeking judicial review, standing is a critical threshold issue. Under the Supreme Court doctrine that has evolved over the past five decades, Article III of the Constitution requires that a plaintiff prove three elements in order to establish standing. First, the plaintiff must show "injury in fact"—that is, that the plaintiff has suffered or will suffer some kind of actual, concrete injury, as opposed to simply a policy disagreement with the government. The idea is that being upset by something that you read in the newspaper is not enough; the government's action must cause you tangible harm of some kind. Second, the plaintiff must show that this injury is "fairly traceable" to the defendant's conduct. This is something like the proximate cause requirement in torts. For instance, some courts have held that citizens living on a river lack standing to sue over discharge that took place far upstream, because the causal connection is too attenuated. Finally, the plaintiff must show that its injury could be remedied, at least in part, by the court. If the court cannot actually do anything to fix the plaintiff's problem, then a judicial victory might make the plaintiff feel better but would have no practical impact. In a long series of cases starting in the 1970s, the Court has struggled to apply this three-part test, with results that do not seem completely consistent and often with sharp divisions among the Justices.

Under this doctrine, a plaintiff seeking to sue to prevent carbon emissions faces some significant challenges. The emissions in question will probably make little or no measurable difference in global temperatures by themselves, and any impact that does exist may be decades in the future. Is this an injury in fact? Also, the harm occurs only because of the combined effect of the emissions at issue with those of thousands or even millions of other emitters. Is the harm "fairly traceable" to the defendant? And even if the court prevents the emissions in question, will that result in any detectable improvement in the plaintiff's future welfare, given the existence of other major carbon sources around the world? It is no wonder that climate standing has been a controversial issue.

The Court confronted this standing issue in *Massachusetts v. EPA*,[36] the suit brought by states, local governments, and environmental organizations discussed earlier in this chapter. Recall that the plaintiffs sought review of EPA's denial of their petition for EPA to regulate greenhouse gas emissions from motor vehicles under the Clean Air Act. A divided panel of the D.C. Circuit ruled in favor of the EPA, in part on the basis of questions about the petitioners' standing. But Justice Stevens, writing for the Court, held that the plaintiffs did have standing (and ruled for them on the merits as well, as we saw earlier).

Before turning to the conventional tripartite test for standing, the Court made four preliminary points. The first was the Court was following congressional instructions to determine the legality of agency conduct. The Court observed that "the parties' dispute turns on the proper construction of a congressional statute, a question eminently suitable to resolution in federal court." The Court also stressed that Congress had authorized such challenges to EPA actions, a significant point because "Congress has the power to define injuries and articulate chains of causation that will give rise to a case or controversy where none existed before." Second, the Court rejected the assertion that injuries are disqualified from serving as a basis for standing merely because they are very widespread. Justice Scalia had long taken the opposing position that no standing exists for widely dispersed injuries. But his view had already been rejected by the Court in other opinions. Third, the Court stressed that the test for standing is easier to meet where a procedural right is involved, in this case, the right to sue over agency action wrongfully withheld. In such cases, the "litigant has standing if there is some possibility that the requested relief will prompt the injury-causing party to reconsider the decision that allegedly harmed the litigant." Fourth and finally, because some of the plaintiffs were state governments, the Court suggested that their standing claim should be treated with particular generosity. Having surrendered some of their sovereign abilities to protect their environments when they entered the union—for example, the ability to negotiate for greenhouse gas reductions with foreign powers—states were now reliant on Congress to help protect their "quasi-sovereign interests."

Of the Court's four general assertions about standing law, the point about the standing of state governments seemed the most novel, though it did have some historical antecedents.[37] But the

---

[36] 549 U.S. 497 (2007).

[37] Notably, some lower courts have recognized some distinctive types of injuries relating to state governments in environmental cases. *See, e.g.,* National Ass'n of Clean Air Agencies v. EPA, 489 F.3d 1221, 1227–1228 (D.C. Cir. 2007) (state agencies have standing when a new EPA rule would allow increased pollution from some

Court's view seems plausible enough. Massachusetts would surely have standing to litigate a claim by a neighboring state that its state line should be moved inwards a foot; it should have similar standing when it is the sea rather than a neighbor attacking its territorial integrity.

With these preliminaries in mind, the Court turned to the tripartite standing test. The first element is injury in fact. As to this element, the Court said, "[t]he harms associated with climate change are serious and well recognized." The Court then referred to a research report, which EPA itself had cited, for a list of possible impacts of climate change: "the global retreat of mountain glaciers, reduction in snow-cover extent, the earlier spring melting of rivers and lakes, [and] the accelerated rate of rise of sea levels during the 20th century relative to the past few thousand years." The Court noted that these effects posed a particular threat to the state's interests: "If sea levels continue to rise as predicted, one Massachusetts official believes that a significant fraction of coastal property will be 'either permanently lost through inundation or temporarily lost through periodic storm surge and flooding events.' " "Remediation costs alone, petitioners allege, could run well into the hundreds of millions of dollars."

As to causation, EPA did "not dispute the existence of a causal connection between man-made greenhouse gas emissions and global warming." EPA did contend, however, that the particular government action that the plaintiffs sought would not have a significant impact, because new automobiles are only one source of greenhouse gases and because the United States as a whole accounts for only a portion of these gases. The Court rejected this "erroneous assumption that a small incremental step, because it is incremental, can never be attacked in a federal judicial forum." Instead, the Court stressed that "[a]gencies, like legislatures, do not generally resolve massive problems in one fell regulatory swoop" but "whittle away at them over time, refining their preferred approach as circumstances change and as they develop a more-nuanced understanding of how best to proceed." Moreover, this particular first step would be far from insignificant: "Considering just emissions from the transportation sector, which represent less than one-third of this country's total carbon dioxide emissions, the United States would still rank as the third-largest emitter of carbon dioxide in the world, outpaced only by the European Union and China."

Finally, the Court was untroubled by the remedial issue. "While it may be true that regulating motor-vehicle emissions will not by

---

sources and thereby make it more difficult for them to establish state implementation plans; court cites *Massachusetts v. EPA* as well as earlier circuit authority).

itself *reverse* global warming, it by no means follows that we lack jurisdiction to decide whether EPA has a duty to take steps to *slow* or *reduce* it." As the Court noted, the government had strongly supported voluntary efforts to reduce greenhouse gases, and it would "presumably not bother with such efforts if it thought emissions reductions would have no discernable impact on future global warming."

Summarizing the Court's holding on standing, Justice Stevens said that, at least according to the uncontested evidence before it, "the rise in sea levels associated with global warming has already harmed and will continue to harm Massachusetts," the risk of catastrophic harm "though remote, is nevertheless real," and the risk would be reduced to some extent if Massachusetts won the case. On the merits, as we saw earlier, the Court then held that EPA had misapplied the Clean Air Act in several critical respects, remanding for further consideration by the agency under the correct statutory standards.

Chief Justice Roberts wrote a vigorous dissent on the standing issue, which was joined by Justices Scalia, Thomas, and Alito. It is worth considering the dissent in detail because it may now have the support of the enlarged conservative majority on the Court. There were four major disagreements between the majority and the dissent. First, the dissent rejected the view that states are entitled to special solicitude in terms of their standing claims. The dissent challenged the majority's assertion of precedential support for this idea. As the dissent pointed out, however, it is unclear whether this "special solicitude" really affected the outcome in the case.

Second, the dissent could find no concrete injury in fact to the state of Massachusetts. In the dissent's view, the "very concept of global warming seems inconsistent with this particularization requirement," since climate change affects the entire human race. After all, the dissent observed, "the redress petitioners seek is focused no more on them than on the public generally"; "it is literally to change the atmosphere around the world." As to the claim that the particularized injury was loss of coast land, that claim failed to meet the demand that injury be "actual or imminent, not conjectural or hypothetical," "real and immediate," and "certainly impending." The dissent could find "nothing in petitioners' 43 standing declarations and accompanying exhibits to support an inference of actual loss of Massachusetts coastal land from 20th century global sea level increases." (Note that, whatever might have been true in 2007, there is now firm evidence that sea level rise is already taking place.) Longer-term injury was alleged in the affidavits, but "accepting a century-long time horizon and a series of compounded estimates renders requirements of imminence and immediacy utterly toothless."

Third, the dissent argued, the Court had misapplied the causation requirement. In the dissent's view, the majority opinion "ignores the complexities of global warming, and does so by now disregarding the 'particularized' injury it relied on in step one, and using the dire nature of global warming itself as a bootstrap for finding causation and redressability." Because the case involved only new vehicles sold in America, the case would affect "only a fraction of 4 percent of global emissions." Furthermore, according to Roberts, predicting future climate change involves a complex web of economic and physical factors, and the plaintiffs "are never able to trace their alleged injuries back through this complex web to the fractional amount of global emissions that might have been limited with EPA standards."

Furthermore, the dissent said, "[r]edressability is even more problematic." Because of projected emissions from China and India, "the domestic emissions at issue here may become an increasingly marginal portion of global emissions, and any decreases produced by petitioners' desired standards are likely to be overwhelmed many times over by emissions increases elsewhere in the world." This point may now have lost some of its punch since those two countries are now committed to controlling their emissions.

Finally, the dissent stressed what it viewed as the constitutional imperatives underlying standing law. Admittedly, when "dealing with legal doctrine phrased in terms of what is 'fairly' traceable or 'likely' to be redressed, it is perhaps not surprising that the matter is subject to some debate." Nevertheless, the Chief Justice said, "in considering how loosely or rigorously to define those adverbs, it is vital to keep in mind the purpose of the inquiry" because the "limitation of the judicial power to cases and controversies 'is crucial in maintaining the tripartite allocation of power set forth in the Constitution.'" This emphasis on the separation-of-powers implications of standing resonates with Justice Scalia's efforts to refocus standing doctrine in a number of earlier cases, but it is telling that the Chief Justice mentioned these concerns only in passing.

The Court did nothing to clarify the issue of climate standing in *American Electric Power v. Connecticut (AEP)*,[38] which involved a lawsuit by several states, New York City, and some nonprofit land trusts against five major electric power companies. The plaintiffs contended that the power companies were committing a public nuisance with their unrestricted use of fossil fuels. Lower court judges had divided about whether suits of this type were even within

---

[38] 564 U.S. 410 (2011).

the jurisdiction of the federal courts. Some district judges said the suits were nonjusticiable "political questions," and the defendants also argued that the injury to the plaintiffs was too diffuse, speculative, and indirect to give them standing to sue. The Supreme Court split evenly on these issues, resulting in no opinions on the subject of jurisdiction. The even-split occurred because Justice Sotomayor had participated in the lower court consideration of the case before she joined the Supreme Court, so she did not participate at the Supreme Court level. But it seems very likely, based on her views in other cases, that she would have found jurisdiction, so at the time there were probably five votes in favor of standing at least in cases where major amounts of carbon are at stake, as they were in *AEP* and in *Massachusetts v. EPA*. How today's Court would rule is unclear.

Lower courts have continued to have difficulty dealing with issues of climate standing, as reflected in *Washington Environmental Council v. Bellon*.[39] The plaintiffs sought to compel a state regulatory agency to regulate greenhouse emissions from five oil refineries under a provision requiring the use of reasonably available control technology (RACT) for some existing sources. The collective annual emissions of these refineries came to about six million tons. The court assumed for purposes of its decision that the plaintiffs would suffer injury-in-fact from climate change, but it concluded that they had failed to show the required causal connection between the refineries and their injuries. The court distinguished *Massachusetts v. EPA* on the ground that the plaintiffs in that case were state agencies, which were entitled to greater judicial solicitude, and more importantly because it considered the much smaller amount of emissions from the five refineries not to constitute a "meaningful contribution" to global emissions, whereas vehicle emissions from the United States as a whole were large enough to register globally.

Although the issue of climate standing presents intriguing theoretical issues, plaintiffs may be able to avoid the problem posed by *Bellon* and similar lower court decisions. Unless a statute requires otherwise—and the citizen suit provision of the Clean Air Act does not—the plaintiff's injury in fact does not have to coincide with plaintiff's legal claim. For instance, if a proposed regulation of mercury from factories would require the installation of scrubbers that would reduce particulates, someone who is injured by the particulates has standing to defend the proposed regulation or demand a stricter regulation. This makes logical sense: the plaintiff has an injury in fact (from the particulates); the injury is directly caused by the lack of scrubbers, and upholding the proposed

---

[39] 732 F.2d 1131 (9th Cir. 2013), *rehearing in banc denied*, 741 F.3d 1075 (2014).

regulation would remedy the plaintiff's problem because it would require scrubbers.

Similarly, if measures taken to lower carbon emissions would also reduce conventional emissions that are harmful to the plaintiff, the plaintiff has standing to demand the lower carbon emissions. In an important case, the D.C. Circuit accepted this theory of standing.[40] Generally it should not be hard to find a plaintiff who suffers the requisite harm from conventional pollutants that are associated with the emission of greenhouse gases. Emissions from coal-fired plants are the perfect example: anything that reduces the use of coal will reduce both carbon emissions and particulate emissions that have serious health effects. So people who breathe the air in question should have standing to demand stricter regulation of coal-fired plants.

Thus far, we have been concerned with the role of the federal courts. Another aspect of the *AEP* case is worth mentioning in connection with this discussion of the judicial role in climate litigation. A number of plaintiffs have brought lawsuits alleging that large-scale carbon emitters were committing a public nuisance that was actionable under the federal common law of nuisance. That federal common law doctrine had been recognized in several cases dealing with interstate water pollution, but the Court later held that this body of federal law evaporated following the passage of the Clean Water Act. Similarly, the Court held in *AEP* that the federal common law of nuisance regarding climate change is displaced by the Clean Air Act's grant of jurisdiction to EPA to regulate greenhouse gases. According to the *AEP* opinion, "*Massachusetts* made plain that emissions of carbon dioxide qualify as air pollution subject to regulation under the Act." The Court continued that "we think it equally plain that the Act "speaks directly" to emissions of carbon dioxide from the defendants' plants." Thus, the Court (notably including Justice Scalia) seemed to have committed itself to recognizing EPA authority over carbon emissions from stationary sources, although that had not been an issue in *Massachusetts v. EPA*.

The court left open the possibility that lawsuits could still be brought under state nuisance law. As discussed below, states and municipal governments have taken up this invitation. So far, their law suits against oil companies have been stalled by a battle over whether the cases can be removed to federal court.

---

[40] *See* WildEarth Guardians v. Jewell, 738 F.3d 298 (D.C. Cir. 2013), in which the court found that non-climate environmental harms of a proposed project were enough to give the plaintiff standing to challenge the agency's determinations on a climate-related issue.

V. Trends in Domestic Climate Litigation

In the years following the decisions in *Massachusetts* and *AEP*, climate litigation has proliferated. Recent domestic climate litigation includes a wide range of cases and claims, but two dominant themes emerge: (1) litigants are refining their approach to common law claims and expanding the scope of claims outward to make greater use of private law claims; and (2) litigants are seeking to situate state obligations to address climate change as a matter of fundamental constitutional and human rights.

Notably, domestic litigants are actively engaging with advanced climate research that helps establish causal links between specific sources of emissions and climate-related harms.[41] In addition, litigants are learning from the early common law cases and are focusing more on state-based common law claims and a wider variety of tort-based claims and causes of action. The combined effect of these trends is that litigants are employing a refined set of causes of actions—drawing from common law claim, consumer protection laws and corporate law claims—and using advanced research to help establish causal connections between defendants' emissions and specific climate-related injuries.

The focus of much of the ongoing litigation is on holding the carbon majors—the largest global fossil fuel companies—legally accountable for their contributions to climate change. Cities and municipalities around the United States, as well as shareholders, investors, and employees, are bringing a flood of claims ranging from nuisance and fraud claims, to new allegations of statutory violations[42] against these large fossil fuel companies.

By July 2022, at least fourteen cities and counties and three states had filed tort suits or launched climate-based fraud investigations against the carbon majors in the United States. The entities bringing suit range from states and large cities such as Rhode Island, Delaware, and Minnesota to New York City, Oakland, and San Francisco to smaller entities such as the cities of Boulder, CO; Santa Cruz, CA; Annapolis, MD; and Charleston, SC. In addition to these public entities, other entities such as crab fisherman in

---

[41] *See* Richard Heede, *Tracing Anthropogenic Carbon Dioxide and Methane Emissions to Fossil Fuel and Cement Producers, 1854–2010*, 122 CLIMATIC CHANGE 229 (2014); FRIEDERIKE OTTO, RACHEL JAMES & MYLES ALLEN, ENVTL. CHANGE INST., THE SCIENCE OF ATTRIBUTING EXTREME WEATHER EVENTS AND ITS POTENTIAL CONTRIBUTION TO ASSESSING LOSS AND DAMAGE ASSOCIATED WITH CLIMATE CHANGE IMPACTS, https://unfccc.int/files/adaptation/workstreams/loss_and_damage/application/pdf/attributingextremeevents.pdf [https://perma.cc/PG8D-2HG6].

[42] *See, e.g.*, Ramirez v. Exxon Mobil Corp., 334 F. Supp. 3d 832, 841 (N.D. Tex. 2018).

California and Oregon have also brought suits against the carbon majors.[43]

The first wave of cities to bring suits against the carbon majors, including New York City, Oakland, and San Francisco, suffered early setbacks. These cases were grounded in arguments asserting that the five largest investor-owned producers of fossil fuels in the world—who, cumulatively are responsible for 11% of global greenhouse gas emissions—have knowingly contributed to climate change, resulting in injuries to the cities due to sea level rise and other climate-induced harms.[44] While New York City sought compensatory damages to cover the costs that the City incurred as a result of climate impacts as well as an equitable order requiring the defendants to abate the nuisance and trespass to which their emissions give rise, San Francisco and Oakland requested more limited relief in the form of an abatement fund to pay for seawalls and other infrastructure needed to address rising sea levels.[45]

These claims were filed in state court, but defendants attempted to remove them to federal court. In both jurisdictions, the plaintiffs cities sought to remand their cases to the state level. In both cases, however, their motions to remand were denied and both cases were dismissed on a number of grounds, including federal preemption and the political question doctrine.[46]

These early decisions demonstrated that the courts were taking climate science seriously even as they struggle to grapple with the appropriate judicial response to a problem of such massive scale. For example, U.S. District Court Judge William Alsup, who presided over the combined Oakland and San Francisco cases, requested a tutorial on climate change science. Following a five-hour tutorial, in his decision to dismiss he declared that "[a]ll parties agree that fossil fuels have led to global warming and ocean rise and will continue to do so," while also calling the scope of the plaintiffs' theory "breathtaking" and ultimately concluding that the "problem deserves a solution on a more vast scale than can be supplied by a district judge or jury in a public nuisance case."[47]

---

[43] Pacific Coast Federation of Fishermen's Associations v. Chevron Corp., 3:18-cv-07477 (ND Cal).

[44] These five are: Chevron Corp, Exxon Mobil Corp, British Petroleum Plc, Royal Dutch Shell, and ConocoPhillips. These defendants are, respectively, the first, second, fourth, sixth and ninth largest cumulative producers of fossil fuels worldwide. *See, e.g.*, Complaint at 1, 2, 31, *New York v. BP P.L.C.*, 325 F. Supp. 3d 466 (S.D.N.Y. 2018).

[45] *See* California v. BP P.L.C., 2018 WL 1064293, at *1 (N.D. Cal. 2018).

[46] *See* City of Oakland v. BP P.L.C., 325 F. Supp. 3d 1017, 1028–29 (N.D. Cal. 2018); California v. BP P.L.C., 2018 WL 1064293, at *5 (N.D. Cal. 2018) (denying plaintiffs' motion for remand).

[47] *City of Oakland*, 325 F. Supp. 3d at 1022.

In a strikingly similar decision dismissing New York City's case, U.S. District Judge John Keenan said "[c]limate change is a fact of life, as is not contested by Defendants. But the serious problems caused thereby are not for the judiciary to ameliorate. Global warming and solutions thereto must be addressed by the two other branches of government."[48]

Despite these early victories for the oil companies, the courts of appeals have largely rejected their claim that the cases belong in federal court.[49] It remains to be seen whether the Supreme Court will intervene. If not, the cases will proceed in state court.

Alongside this evolving body of common law-based litigation, there are growing efforts to use corporate law to compel fossil fuel entities to disclose information and modify their business practices. Together, the common law and corporate law litigation is bringing the carbon majors under increased legal, ethical, and financial scrutiny.

In addition to the ongoing common law and corporate law claims, litigants are drawing upon constitutional and human rights law to bring claims arguing that the state has a fundamental legal obligation to address climate change.[50] These cases are moving beyond the constraints of existing statutory and regulatory regimes to try to situate state obligations to address climate change as a matter of fundamental constitutional and human rights.

The case that best embodies this approach in the United States is *Juliana v. United States*.[51] *Juliana* began in 2015, when twenty-one young people filed suit against the United States claiming that the federal government had violated their legal rights by knowingly contributing to climate change and failing to reduce domestic greenhouse gas emissions or otherwise address the causes and consequences of climate change.

More specifically, the youth plaintiffs alleged that the government has deprived them of their right to a safe climate without due process of law and thereby violated their constitutional rights of due process and equal protection. The plaintiffs also alleged that the

[48] *City of New York v. BP P.L.C.*, 325 F. Supp. 3d 466, 474–75 (S.D.N.Y. 2018).

[49] *See, e.g.*, Mayor & City Council of Baltimore v. BP P.L.C., 31 F.4th 178 (4th Cir. 2022); Board of County Commissioners of Boulder County v. Suncor Energy (U.S.A.) Inc., 25 F.4th 1238 (10th Cir. 2022); City of Oakland v. BP PLC, 969 F.3d 895, 907 (9th Cir. 2020) ("Oakland"), cert. denied, 141 S. Ct. 2776 (2021).

[50] *See, e.g.*, Rb.'s-Gravenhage 24 juni 2015, AB 2015, 336 m.nt. Ch.W. Backes (Stichting Urgenda/Staat der Nederlanden).

[51] *See Juliana v. United States*, 217 F. Supp. 3d 1224, 1233 (D. Or. 2016). The complement to the *Juliana* case in the United States is the *Urgenda* case in the Netherlands, which was much more successful. Urgenda v. The Netherlands (Supreme Court, 20 December 2019) ECLI:NL:HR:2019:2007.

federal government is the sovereign trustee of the "country's life-sustaining climate system" and has failed in its duty of care to present and future generations to "take affirmative steps to protect those trust resources."[52] The core of the youths' claims was that the government's actions and inactions "have so profoundly damaged our home planet that they threaten Plaintiffs' fundamental constitutional rights to life and liberty."[53]

The *Juliana* litigation pushes against the edges of constitutional law jurisprudence and seeks bold new declarations of federal failure and federal power. At the core of the plaintiffs' claims is the notion that climate change poses an immediate and existential threat to present and future generations and that the government must, in some meaningful way, be accountable for addressing this threat. Federal District Court Judge Anne Aikens affirmed this core assertion and responded to the youths' claims by acknowledging the far-reaching failures of domestic environmental law and finding that "the right to a climate system capable of sustaining human life is fundamental to a free and ordered society."[54] This was an unprecedented finding.

On appeal, the Ninth Circuit in a 2:1 decision ordered the district court to dismiss the case on the basis of standing.[55] While articulated in terms of standing law, the majority's concerns seemed more focused on the separation of powers.

The court found that two of the three requirements for standing were satisfied. In terms of injury-in-fact, the injuries were sufficiently particularized. One plaintiff, for instance, "alleged she was forced to leave her home because of water scarcity, separating her from relatives on the Navajo Reservation." The court also found that the plaintiffs had satisfied the causation element of standing:

> The plaintiffs' alleged injuries are caused by carbon emissions from fossil fuel production, extraction, and transportation. A significant portion of those emissions occur in this country; the United States accounted for over 25% of worldwide emissions from 1850 to 2012, and currently accounts for about 15%. And, the plaintiffs' evidence shows that federal subsidies and leases have increased those emissions. About 25% of fossil fuels extracted in the United States come from federal waters

---

[52] First Amended Complaint for Declaratory & Injunctive Relief at 92–93, Juliana v. United States, 217 F. Supp. 3d 1224 (D. Or. 2015).

[53] *Id.* at 1261 n.28.

[54] Juliana v. United States, 217 F. Supp. 3d 1224, 1250 (D. Or. 2015).

[55] *Juliana* v. United States, 947 F.3d 1159 (9th Cir. 2020).

and lands, an activity that requires authorization from the federal government.[56]

The majority concluded, however, that the third element of standing was not satisfied because the injury they alleged could not be remedied by a court. The Ninth Circuit held that the remedy sought by the plaintiffs—an order requiring the government to develop a plan to "phase out fossil fuel emissions and draw down excess atmospheric CO_2"—was beyond the constitutional power of the court.[57] In so doing, however, the court affirmed the severity of the climate crisis, finding that the "[c]opious expert evidence establishes that this unprecedented rise stems from fossil fuel combustion and will wreak havoc on the Earth's climate if unchecked."[58]

Even as it denied its own ability to contend with the impending crisis, the Ninth Circuit warned that "it will be increasingly difficult . . . for the political branches to deny that climate change is occurring, that the government has had a role in causing it, and that our elected officials have a moral responsibility to seek solutions."[59]

It seems exceedingly unlikely that the current Supreme Court would rule in favor of a lawsuit such as *Juliana*. Given that this is the most conservative Court in nearly a century, a ruling in favor of the plaintiffs would require that two of the conservative Justices (perhaps Roberts and Kavanaugh) overcome concerns about the separation of powers. They would also have to accept the idea of a constitutional right to a safe climate, despite having adopted a very narrow view of unwritten constitutional rights when discarding the long-established right to an abortion. Nothing is impossible, but this seems to come close.

Although they are by no means slam dunks, the cases brought by state and local governments against oil companies seems to have better prospects of success. As we will see in the next chapter, these lawsuits are only the tip of the iceberg in terms of state climate change initiatives.

Further Readings

Cinnamon P. Carlarne, *U.S. Climate Change Law: A Decade of Flux and an Uncertain Future*, 69 AM. U. L. REV. 387 (2019).

---

[56] *Id.* at 1169.

[57] *Id.* at 1172.

[58] *Id.* at 1166 (the court stated that "[a]bsent some action, the destabilizing climate will bury cities, spawn life-threatening natural disasters, and jeopardize critical food and water supplies").

[59] *Id.* at 1175.

Ann E. Carlson and Megan M. Herzog, *Text in Context: The Fate of Emergent Climate Regulation After* UARG *and* EME Homer, 39 HARV. ENV. L. REV. 23 (2015).

Megan Ceronsky and Tomás Carbonell, *Section 111(d) and the Clean Power Plan: The Legal Foundation for Strong, Flexible, and Cost-Effective Carbon Pollution Standards for Existing Power Plants*, 44 ENVTL. L. REP. NEWS & ANALYSIS 11086 (2014).

Daniel A. Farber, *A Place-Based Theory of Standing*, 55 UCLA L. REV. 1505 (2008).

J.B. Ruhl, *Climate Change and the Endangered Species Act: Building Bridges to the No-Analog Future*, 88 B.U. L. REV. 1, 6 (2008).

Joanna Setzer & Rebecca Byrnes, *Global Trends in Climate Litigation: 2020 Snapshot*, Grantham Research Institute on Climate Change and the Environment (2020).

Robert Sussman, *Power Plant Regulation Under the Clean Air Act: A Breakthrough Moment for U.S. Climate Policy?*, 32 VA. ENV. L. REV. 97 (2014).

Paul Weiland, Robert Horton, and Erick Beck, *Environmental Impact Review*, in MICHAEL B. GERARD AND JODY FREEMAN, GLOBAL CLIMATE CHANGE AND U.S. LAW (2d ed. 2014).

Chapter 7

FEDERALISM AND CLIMATE CHANGE

In ratifying the U.N. Framework Convention on Climate Change (UNFCCC), the United States committed itself to regulating greenhouse gases. Two provisions of the UNFCCC applicable to the United States are particularly relevant. First, in Article 3(3) of the Convention, the parties pledge to take "precautionary measures to anticipate, prevent or minimize the causes of climate change and mitigate its adverse effects." Second, under Article 4(1)(b), the parties agree to "formulate, implement, publish and regularly update . . . measures to mitigate climate change by addressing anthropogenic emissions by sources." In particular, developed countries such as the United States commit to adopting "national policies and tak[ing] corresponding measures on the mitigation of climate change, by limiting its anthropogenic emissions of greenhouse gases and protecting and enhancing its greenhouse gas sinks and reservoirs." One might have expected the United States government to turn to the task of implementing these responsibilities through domestic regulation. But as it turned out, climate mitigation efforts in the United States took a different turn.

The history of climate change mitigation in the United States contains what seems to be a paradox. Because it is a global problem, climate change seems like a natural subject for the attention of the federal government, but an unlikely subject for state and local activism. Yet state and local governments were the first to enter the field. The Bush Administration failed to address climate change either through administrative action or legislation. Congress tried but failed to pass comprehensive climate litigation in 2010, at a time when several states had already created cap-and-trade schemes. Indeed, when the federal government finally did begin to regulate carbon emissions under the Clean Air Act, it did so only because of lawsuits supported by state governments. During the Obama Administration, it seemed that EPA might finally be catching up to the leading states through the Clean Power Plan discussed in the previous chapter. The Trump Administration reversed course completely, leaving the Biden Administration to try to pick up the pieces. Thus, the federal government's efforts to cut emissions have been erratic.

The issues in this area reflect the nature of the American system of federalism. By default, states have plenary power to make policy, subject only to the limitations of federal law. Thus, in the absence of some specific constitutional barrier, states are the default policymakers. Over time, the scope of federal legislation has dramatically increased, but state law remains central in many domains such as land use planning, and states actively participate in all spheres of domestic policy. Yet the actions of one state may impinge on national policy or on the people and economies of other states. Federalism limits on the states may stem directly from the Constitution itself and the judicial doctrines elaborating the Constitution's meaning. Or they may stem from federal legislation and the Supremacy Clause, which designates federal statutes as part of the "supreme law of the land." The federal courts referee these disputes over the division of authority between different states and between states and the federal government.

In this chapter, we consider the role that state and local governments have played in addressing climate change (with the main emphasis being on state governments). Part I provides an overview of state programs. The positions of states regarding climate change could hardly be more diverse. Some states are doing little to limit emissions and are fighting bitterly against EPA's efforts, while others are on the forefront of climate action. Still, there does seem to be broad state support for at least making an effort to encourage renewable energy, and some states have been aggressive in their efforts to address climate change. Politically liberal states have been in the forefront, but conservative states have also built up their renewable resources. For instance, Texas has been aggressive in promoting wind power. As a result, it has by far the highest wind generation capacity in the country, followed by Iowa and Oklahoma, with Kansas fourth.

After this background on state activities, the remainder of the chapter focuses on the legal challenges that states may face when they do try to limit emissions. Part II is devoted to the dormant commerce clause, a constitutional doctrine that can pose a particular problem for state energy regulation. In particular, state governments may be confronted by claims that their programs discriminate against out-of-state firms or may be an indirect effort to extend the states regulatory authority beyond its own borders. As one example, state renewable energy programs can be attacked because of their repercussions on energy producers and consumers in other states.

Part III considers other legal issues that may hamper state efforts. Those challenges can take a variety of forms. Industry may argue that federal statutes regulating energy preempt state emissions limitations or state efforts to promote renewables. A related argument is that state regulatory efforts impinge on federal authority over foreign affairs because of the international nature of the climate problem. Finally, industry may argue that cooperative efforts between states and other states, or between states and foreign jurisdictions, are unconstitutional in the absence of congressional approval.

There is obviously plenty to keep lawyers busy in designing state programs to minimize legal risks and in litigating the validity of the programs. So far, states have done fairly well in litigation, but the law is still in the process of developing. It remains to be seen how the doctrine will eventually shake out. In the meantime, an understanding of the areas in dispute remains crucial.

I. State and Local Climate Change Programs

American state governments have actively engaged with climate change on many fronts. States have also been active in promoting renewable energy, sometimes under the climate umbrella and sometimes independently. Every state has adopted some programs that either directly address climate change or encourages renewable energy and energy efficiency.

About three-quarters of the states have state climate action plans of some kind. About half the states have also set targets for reducing emissions. State targets vary widely, with perhaps the most ambitious being Oregon's 2050 target of reducing emissions from all sources by 75 percent from the 1990 level and California's 80 percent target for that year. Notably, states remained active even under the Bush and Trump Administrations, when the federal government steadfastly refused to take any action at all to reduce emissions, and even seemed dedicated to increasing them through expanded production of fossil fuels.

Many states have adopted renewable portfolio standards (RPS), which require that a certain percentage of retail electricity sales be derived from renewable sources. As of 2022, thirty-eight states have established an RPS or a renewable goal. Twelve of those states mandate the use of 100 percent clean electricity by 2050 or earlier. California's RPS requires sixty percent renewables by 2030 and 100 percent by 2045. New York's targets are seventy percent by 2030 and 100 percent by 2040.

There are significant variations in these standards from state to state. The most obvious are differences in the percentage of renewables required. But there are also differences in what energy sources count as renewables, with nuclear power and hydropower included in some places but not others. There may also be differences in whether all utilities are covered or only investor-owned activities, and whether there are specific requirements for particular sources such as solar or wind, and in the deadlines. Finally, there may be differences with respect to the tradability of renewable credits. For example, in some states, a firm that gets more of its energy from renewables than required may be able to sell credits to a firm that is under the target.

There is a political tilt in terms of adoption of RPSs, with Democratic-leaning states the most likely to have binding standards. But this is only a trend, and there are some significant exceptions. Nevertheless, there are two areas of the country that have been holdouts from RPS requirements. One, in the West, is a strip consisting of Idaho, Wyoming, and Nebraska. (In fact, Wyoming almost passed a "reverse RPS" in 2017 that would have penalized companies for using renewables.) The other exception is the Southeastern United States, where the only states that do have some type of RPS are South Carolina, North Carolina, and Virginia (and only the latter two are binding). At present, the political equilibrium seems to be stable, with little movement in either direction. There has been an effort recently on the part of conservatives to roll back RPSs, but so far this effort has met only very limited success.

Among the states that are addressing climate change in meaningful ways, California has played a leading role. California legislation focusing specifically on climate change dates back to a 1988 law mandating an inventory of California greenhouse gas emissions.[1] In 2006, Governor Arnold Schwarzenegger signed the California Global Warming Solutions Act, usually referred to as AB 32,[2] which requires California to reduce emissions to the 1990 level by 2020. This California law generated worldwide attention, including enthusiastic approval by the British Prime Minister at the time it was passed. The California effort undoubtedly received additional attention because the governor was an international celebrity and because it was such a stark contrast with the Bush Administration's recalcitrance. But there were also more tangible international steps involving California. The Prime Minister and the Governor of California entered into an agreement to share best practices on market-based systems and to cooperate to investigate new technologies; similar agreements now exist between California and states and provinces in Australia and Canada. (We will discuss possible legal issues relating to such agreements later in the chapter). California has also pursued discussions with governmental authorities in China.

California has implemented AB 32 aggressively. The law itself is notably brief and gives the government enormous discretion about how to achieve its goals, though it does rule out a carbon tax. The California Air Resources Board (CARB) first developed nine "early

---

1 AB 4420 (Sher), Chapter 1506, Statutes of 1988.

2 AB 32 (Nunez), Chapter 488, California Statutes of 2006, codified at CAL. HEALTH & SAFETY CODE §§ 38500 *et seq.*

action" measures, some of which focus on reducing emissions of non-CO_2 greenhouse gases. Another important early action was a low-carbon fuel standard to reduce the carbon intensity of transportation fuels by 10 percent by 2020. But the CARB's most important action was to establish an emissions trading system, which was discussed in more detail in chapter 4. California's cap-and-trade program sets a declining, statewide cap on greenhouse gas emission. The program originally covered about 600 industrial facilities, with fuel distributors having been added to the program more recently. Many allowances have been distributed free to firms, but an increasing percentage will be auctioned. Since the program began, auctions have generated $12 billion for the state, which has been spent on climate-related programs. In 2017, the state extended the life of the trading system through 2030, with a target of reducing its total emissions thirty percent below 1990 levels.

Another notable effort is the Northeast Regional Greenhouse Gas Initiative (RGGI). RGGI, which is currently composed of eleven states, created a multistate trading system for power plant emissions with the goal of achieving a 10 percent reduction by 2019.[3] In 2013, the cap was reset to 91 million tons of carbon, down from 165 million tons. A quarter of the proceeds are auctioned, with the proceeds going to finance energy efficiency programs or reduce fee hikes caused by the program. Indeed, many of the carbon reductions associated with the program have stemmed from these energy efficiency programs rather than from the cap itself. The allowance prices remain low, indicating that the cap is still generous, but the cap is set to decline by 2.5 percent annually. In mid-2017, the RGGI states tightened the cap in response to the Trump Administration's climate policies, which will result in a thirty percent cut in emissions from 2020 to 2030.

In addition to actions at the state level, many cities have adopted climate action plans. Although cities do not have the same extensive regulatory powers as state governments, some specific aspects of emission reductions relate to municipal activities in a fairly direct way. Efforts by city governments have taken many forms. Urban planning and land use control is an important municipal function with important implications for climate change. For instance, cities may use their building codes to encourage more energy-efficient buildings and their transit planning to promote public transportation. One area of interest is promotion of transportation-oriented development, where the goal is to promote additional development close to public transportation hubs. Cities

[3] *See* https://www.rggi.org.

can also reduce barriers to the use of renewable energy, such as zoning restrictions that could hinder rooftop solar.

In addition, city governments can reduce their own energy use and can adopt renewable sources of energy, such as generating electricity from methane produced by waste. Municipalities own a large number of buildings and vehicles such as police cars, so potential emissions reductions are not trivial. Finally, some cities run their own municipal electrical utilities, which sometimes have adopted ambitious renewable energy and energy efficiency programs. Given the proportion of the population and the economy found in urban areas, these are not insignificant steps.

Apart from more formal cooperative efforts between state and local jurisdictions, which are discussed later in the chapter, extensive networks and collaborative groups have also arisen. These groups provide forums for mutual encouragement and sharing of expertise. They also help identify best practices and promising new techniques. In many cases, these efforts span national borders, with active participation from sub-national governments around the world.

Before we turn to the legal issues raised by state and local climate programs, it is worth taking a moment to consider why states and localities have entered this area at all. Their activity in the climate area seems somewhat surprising in light of their normal responsibilities. State governments are predominantly concerned with the day-to-day concerns of their citizens and with traditional areas of law such as criminal law, family law, and contracts. International problems like climate change seem well outside these normal concerns.

It is natural to ask why states have chosen to take action on what is, after all, a global problem. Given the need for global cooperation to reduce emissions, it seems odd that a relatively small entity such as a U.S. state would be willing to invest in cutting emissions on its own. The costs of the emission reduction are felt within the state, but almost all of the climate reduction benefits go to outsiders, including many outside the United States entirely. No single state can make more than a tiny contribution to the global effort to reduce global emissions. This seems to be a classic example of the tragedy of the commons or a prisoner's dilemma. Thus, the voluntary activism of many states and cities is puzzling.

No doubt each state presents its own complicated political story. But there are some general factors that seem to be at work. One is that actions that reduce carbon emissions generally reduce other kinds of emissions as well, so they have immediate benefits to states with air pollution problems. These co-benefits may justify the state programs even without consideration of climate change. In this way, states can reap benefits without waiting for global action on climate change. Moreover, states may see economic advantages in diversifying their sources of power, which is one reason why Texas has done so much to develop wind power so as to avoid complete dependence on natural gas. States may see advantages in expanding a local industry, such as solar installation, that involve many jobs. States may also hope that leadership on climate policy may promote innovation hubs in clean energy, ultimately bringing new investment to the state. Some states may also have seen potential advantages to acting on climate change ahead of the federal government in order to position themselves to comply with any new federal requirements. And of course, climate change is an issue that concerns many members of the public, not to mention some political leaders themselves.

Whatever the reason, it is clear that many American states and cities have decided to take action against climate change. Our present concern is the legal validity of those actions. It is difficult to say anything in general about the legality of municipal actions under state law, because local government law varies so much between states. The traditional rule is that municipalities have only the powers delegated to them by state governments, although some now have home rule powers. State governments, too, may be limited by their own state constitutions. These issues of state law, however, are beyond the scope of this book.

Regardless of their powers under state law, however, cities and states must always comply with federal law. The general rule in American law is that state governments have the power to do anything that is not prohibited by federal law. We begin our consideration of these federal limitations with one of the main restrictions on state and local regulation, the dormant commerce clause.

II. Climate Regulations and the Dormant Commerce Clause

The otherwise-broad powers of state governments are limited by the dormant commerce clause, a constitutional doctrine created by the Supreme Court under the aegis of the Commerce Clause. The Commerce Clause itself is simply a delegation of power to Congress to regulate interstate commerce. But the Supreme Court has viewed the clause as implicitly limiting state interference with interstate commerce. By protecting interstate commerce from state interference, the Court has attempted to prevent balkanization of the U.S. economy and state discrimination against outsiders, who are not represented in the state's political process. But given that almost any regulation has interstate ramifications in our highly integrated national economy, line drawing has proved difficult.

Over two centuries, the Court's decisions have created what is now known as the dormant commerce clause doctrine ("dormant" because it does not involve the exercise of Congress's authority under the Commerce Clause.) Some scholars and judges criticize the Court's efforts for having a weak foundation in the constitutional text. The Supreme Court has inferred limitations on state authority from the clause since the early Nineteenth Century, although both the precise rationale and the Court's test have varied over time. But whatever might be said about these issues as a matter of constitutional theory, it seems clear that a majority of Justices support the doctrine. As a practical matter, regardless of theory, state regulators and regulated parties must operate under the requirements of the doctrine.

The Supreme Court has distinguished between regulations that merely burden interstate commerce and those that discriminate against it or regulate commerce outside of the state. Discriminatory and extraterritorial regulations face more stringent scrutiny, and for that reason we will focus on those categories. Regulations that merely burden interstate commerce also face scrutiny, however, although in a less rigorous mode. Under the balancing test of *Pike v. Bruce Church, Inc.*,[4] state laws burdening commerce are invalid if the burden on interstate commerce is "clearly excessive in relation to the putative local benefits."[5]

Although defending climate regulations under the *Pike* test requires extensive factual inquiry, states should be well positioned to win those disputes. They should be able to demonstrate the strength of their interest in reducing emissions, given what we know about climate change. In *Massachusetts v. EPA*,[6] the Court emphasized that climate change threatens the state's semi-sovereign interest in the welfare of its citizens and in protecting its territory (as a result of sea level rise). In general, states seem to fare well under the balancing test, regardless of the type of regulation under attack. They can be expected to do well under the *Pike* test in the domain of climate regulation as well, although no doubt there will also be some losses. But if regulations are classified as discriminatory or extraterritorial, their fate is likely to be grim.

Determining the limits of state regulatory authority is especially challenging in the realm of electricity regulation. The nation's energy system is tightly integrated. With the exception of much of Texas, the

---

4 397 U.S. 137 (1970).

5 *Id.* at 142.

6 549 U.S. 497 (2007).

48 contiguous states are part of an interconnected electrical transmission system. The market for fossil fuels also extends well beyond state boundaries, being national if not international in scope. It is not surprising that climate regulations, which often address the energy sector, are prone to challenge under the dormant commerce clause.

Climate regulations are also apt to raise issues under the Commerce Clause because of efforts by regulators to combat leakage. The usefulness of a state's restrictions on emissions can be severely undercut if the effect is to shift carbon emissions out of state rather than eliminating them. For instance, as discussed in Chapter 4, a state's effort to limit emissions from its own products may have little effect if its citizens merely shift their buying to cheaper, out-of-state sources with high emissions. Efforts to impede leakage have clear interstate implications, since they are aimed at preventing state regulations from increasing emissions outside the state. This makes them an especially likely target for attack under the dormant commerce clause.

Dormant commerce clause challenges are not merely a theoretical threat. For instance, in *Rocky Mountain Farmers Union v. Goldstene*,[7] a federal district court struck down California's low carbon fuel standard. The standard attempted to measure the carbon intensity of vehicle fuels based on a lifecycle analysis from production to combustion. The court found the standard to be discriminatory because it included geographic factors such as transportation distances and the carbon-intensity of the electricity used for production from the local grid.[8] These factors disfavored some out-of-state producers, especially from the Midwest where coal is a favored fuel for generating the electricity used to process corn into ethanol. The court also found that the standard was impermissibly extra-territorial because it took into account carbon emissions that occurred outside of the state.[9] In effect, the district court thought, California was trying to regulate emissions outside of its own borders that were beyond its jurisdiction.

We will discuss the appellate court's opinion and the ultimate fate of the regulations later. But the district court's ruling highlights the importance of careful attention to possible claims of discrimination and extraterritoriality by state regulators. We will focus first on claims that laws discriminate against interstate commerce, and then turn to claims that they amount to extraterritorial regulation of emission sources.

---

7 843 F. Supp.2d 1071 (E.D.Cal. 2011).

8 *Id.* at 1086–1089.

9 *Id.* at 1090–1093.

A. The Ban on Discriminatory State Regulation

If a state attempts to control the carbon intensity of goods and services sold within the state, out-of-state producers may well complain that their goods are being subject to standards that are more restrictive than their home states. Those out-of-state firms are likely to claim that the state is discriminating against interstate commerce. For instance, coal-fired power plants located outside of the state may protest measures that make their electricity less attractive to in-state utilities. They are likely to argue that the regulations discriminate against them versus renewable energy generators within the state.

The Supreme Court has taken vigorous action against what it perceives to be discriminatory state laws. The seminal modern case is *City of Philadelphia v. New Jersey*,[10] in which New Jersey limited imports of solid waste due to concerns about limited landfill capacity. Bypassing disputes about the purpose and effect of the program, the Court emphasized the principle of non-discrimination: "[W]hatever New Jersey's ultimate purpose, it may not be accomplished by discriminating against articles of commerce coming from outside the State unless there is some reason, apart from their origin, to treat them differently."[11] Having found that the statute before it was discriminatory, the Court roundly condemned this "attempt by one State to isolate itself from a problem common to many by erecting a barrier against the movement of interstate trade."[12]

A state law is discriminatory if it explicitly targets out-of-state firms for negative treatment or if it was clearly intended to do so. But it can also be classified as discriminatory because of its effects. A state law "discriminates only when it discriminates between similarly situated in-state and out-of-state interests."[13] A key questions is when two producer are "similarly situated." High-carbon intensity and low-carbon intensity producers might or might not be considered "similarly situated" under this approach.

Moreover, to render a state law discriminatory, it is not enough that some out-of-state firms are disadvantaged or that the state's regulations modify the "natural" operation of the market. Under *Exxon Corp. v. Maryland*,[14] interfering with the natural operation of the national market does not in itself implicate the Commerce Clause; instead, plaintiffs must show that the regulation will

---

10 437 U.S. 617 (1978).

11 *Id.* at 627.

12 *Id.* at 628.

13 Allstate Ins. Co. v. Abbott, 495 F.3d 151, 163 (9th Cir. 2007).

14 437 U.S. 117 (1978).

increase the total market share of in-state goods.[15] Even if climate regulations impact some imported goods more than some local goods, the argument that a regulation that discriminates in effect (rather than on its face) should be viewed cautiously. The courts have cautioned that "[t]he proof of the pudding here must be in the eating, not in the picture on the box as seen through the partial eyes of the beholder—which is especially true in a case where neither facial economic discrimination nor improper purpose is an issue."[16]

A court of appeals case dealing with electricity transmission helps flesh out the meaning of discrimination in this context.[17] The case involved a Minnesota law that granted a right of first refusal to companies owning transmission projects in the state for new projects that would connect to their lines. The incumbent transmission owners are generally but not always local utilities. These laws could stifle competition and discourage new long-distance transmission projects needed to connect renewables with their markets.

An out-of-state transmission company challenged the right of first refusal as unconstitutionally discriminatory against out of state companies. The Eighth Circuit was unconvinced. The court emphasized that some out-of-state firms had a right of first refusal, and some in-state firms lacked it. Moreover, in the court's view, the state law was not primarily aimed at protecting local utilities. Instead, its central purpose was to maintain the benefits of its system of electricity regulation. The court might be criticized for being naïve about the effects of the Minnesota law, but it was right to have looked for a discriminatory effect.

A recurring issue is posed by state laws favoring products including electricity whose production involves low or zero emissions. Out-of-state firms are likely to argue that their product is identical to that of other firms even if the production process involves higher carbon emissions. In response, regulating states will argue that subjecting imports to the same standards as home-produced products should not be considered discriminatory regardless of how carbon intensities are distributed between local and imported goods.[18] In the eyes of regulators, differing carbon intensities reflect real differences among the goods consumed in the state and real differences between the harm they inflict within the state through their contribution to climate change.

---

15 *Id.* at 126 n. 16.

16 Black Star Farms LLC v. Oliver, 600 F.3d 1225, 1232 (9th Cir. 2010).

17 LSP Transmission Holding LLC v. Siegen, 954 F.3d 1019 (8th Cir. 2020).

18 *See* Kirsten H. Engel, *The Dormant Commerce Clause Threat to Market-Based Environmental Regulation: The Case of Electricity Deregulation*, 26 ECOLOGY L.Q. 243, 288–289 (1999).

On this basis, the Ninth Circuit overturned the district court opinion discussed earlier and rejected a discrimination claim against the California low carbon fuel standard (LCFS) in *Rocky Mountain Farmers Union v. Corey*.[19] Location is obviously a relevant factor in the lifecycle analysis of fuels. For instance, fuels that are produced further away will result in more transportation emissions on their way to market. Moreover, electricity is used to produce biofuels such as corn ethanol, and the amount of carbon produced by electricity generation varies greatly between states, depending in part on how much coal is used. Thus, if "discrimination" simply means taking the location of production into account, the LCFS was certainly guilty.

Nevertheless, the majority on the Ninth Circuit held that the LCFS was nondiscriminatory. The court faulted the trial judge for "ignoring GHG emissions related to: (1) the electricity used to power the conversion process, (2) the efficiency of the ethanol plant, and (3) the transportation of the feedstock, ethanol, and co-products," because "those factors contribute to the actual GHG emissions from every ethanol pathway, even if the size of their contribution is correlated with their location." According to the appellate court, "California, if it is to have any chance to curtail GHG emissions, must be able to consider all factors that cause those emissions when it assesses alternative fuels." Thus, the court added, "[t]hese factors are not discriminatory because they reflect the reality of assessing and attempting to limit GHG emissions from ethanol production."[20]

Like the majority, the dissenting judge also rejected the industry argument that use of life-cycle analysis was inherently impermissible. The dissent concluded on narrower grounds, however, that the LCFS was discriminatory on its face and that it was not narrowly tailored to the state's legitimate interest in reducing carbon emissions. The reason related to the details of the LCFS. The regulation provided default carbon intensities for various categories of producers such as Midwest producers of corn ethanol. Producers could then present arguments for modifying the default values due to their circumstances. The dissenting judge argued that the default values were too crude a tool. Instead, the state needed to apply lifecycle analysis on an individualized basis to each producer. By doing so, the state would have narrowly tailored the regulation to the point where it would be constitutional. The majority, however, considered the default values a reasonable way of making the regulation more efficient.

It is not surprising that courts have found this kind of problem difficult. Discrimination can be a slippery concept in any setting, and

---

[19] 730 F.3d 1070, 1077 (9th Cir. 2013) cert. denied, 573 U.S. 946 (2014).

[20] *Id.* at 1092–1093.

the Commerce Clause is no exception. The ban on regulations that discriminate against commerce in effect is particularly poorly defined. A court that is unsympathetic to state climate regulation can probably find some doctrinal basis for objecting to efforts to control carbon leakage. But if courts understand the reasons for state regulation, well-designed regulations should be able to avoid the anti-discrimination rule. The crucial question is whether other courts will follow the Ninth Circuit in recognizing that carbon intensity is a legitimate, non-discriminatory basis for regulation.

B. The Extraterritorial Regulation Issue

A related issue is whether measures designed to control leakage violate the dormant commerce clause's strictures against "extraterritorial" regulation. It is clear, of course, that states cannot punish conduct that lacks any connection to the state. A crime committed in Ohio cannot be prosecuted by California. But in the cases we are about to consider, the state does have some connection with the conduct, because the regulation applies directly to sale or use of a product inside the state. Nevertheless, the local regulation may effectively control conduct outside of the state.

In terms of the dormant commerce clause, the claim is that such regulations are nevertheless extra-territorial because the regulation takes into account emissions outside the state.

From the view of the challengers of state carbon regulations, the analogy would be to a state banning the import of products unless the producers had met the state's labor regulations. If states were allowed to do that, either national markets would be hopelessly fractured or states with large numbers of consumers would be able to control labor policy across the nation. But climate regulators would argue that this is a false analogy. The labor conditions under which products are produced in other states have no direct effect in California even if the products are later sold there. The only harm to California when those goods are imported is economic because the goods are cheaper and may outcompete California products. But carbon emissions from production in other states in connection with California sales do directly impact California, so California has an interest other than protecting local industry.

Unfortunately, the Supreme Court has failed to provide a clear test for determining when a regulation is extra-territorial. The extraterritoriality doctrine stems from a handful of Supreme Court cases. None of those cases have clearly explained when a regulation's effects on conduct outside the jurisdiction constitute "extraterritorial regulations." The cases also deal with situations far removed from environmental regulation. Nevertheless, it behooves us to examine the handful of Supreme Court opinions applying the doctrine.

In one of the key modern cases, the Court struck down a state law requiring liquor wholesalers to give "most favored nation" treatment to New York retailers—in other words, wholesalers had to charge New York retailers the lowest price they offered anywhere else. The New York law created powerful pressure on wholesalers to avoid cutting prices in other states, because doing so meant they would also have to cut prices in New York, doubling the loss of revenue. Thus, the New York law indirectly penalized wholesalers for price cuts in surrounding states, something the state had no right to do directly. A few years later, the Court struck down a similar law from another state.[21] These cases relied on a Depression-era case, *Baldwin v. G.A.F. Seelig, Inc.*,[22] which involved an effort to protect New York dairy farmers from out-of-state competition. In effect, buyers had to pay the New York price no matter where they purchased the milk, if they wanted to resell the milk in New York.[23] The *Baldwin* Court held that the state could not extend its price control regime beyond its borders.

Because the few Supreme Court cases on the subject have failed to give clear guidance, the lower courts have found it difficult to interpret the extraterritoriality doctrine. On the whole, they have resisted efforts to recast the limited Supreme Court's precedents into a broad shield against state regulation. In the few cases to find extraterritoriality, the courts of appeals considered the state law to be the functional equivalent of direct penalties on out-of-state actors or conduct. In contrast, lower courts have refused to find extraterritoriality when a regulation does not directly regulate the actions of parties located in other states and instead regulates contractual relationships with at least one in-state party. Some lower courts have limited the extraterritoriality doctrine to cases involving price regulations, hewing closely to the facts of the Supreme Court cases.

The most important case to date to consider extraterritoriality in the climate change context is *Rocky Mountain Farmers*. In addition to the argument that the LCFS discriminated against interstate commerce, industry also argued that it was extraterritorial because it penalized producers for carbon emitted outside of California. As mentioned earlier, the district court had accepted this argument. The Ninth Circuit rejected this argument because the LCFS "says nothing at all about ethanol produced, sold, and used outside California, it does not require other jurisdictions to

---

[21] The details of the laws differed, but the Court found no practical difference between the New York and Connecticut laws, which were equally coercive of out-of-state conduct. *Id.* at 338–399.

[22] 294 U.S. 511 (1935).

[23] *Id.* at 520.

adopt reciprocal standards before their ethanol can be sold in California, it makes no effort to ensure the price of ethanol is lower in California than in other states, and it imposes no civil or criminal penalties on non-compliant transactions completed wholly out of state." Although it was true that the LCFS did reward the use of lower carbon methods of production for out-of-state producers, encouragement is different than control. Finally, the LCFS was based on differences in fuels that had a direct relationship to harm done within the state. Thus, the extraterritoriality doctrine was inapplicable.

In another recent case, the Tenth Circuit rejected an extraterritoriality challenge to Colorado's renewable portfolio standard. The court argued that the doctrine should be limited to cases involving "price control or price affirmation statutes that link in-state prices with those charged elsewhere and discriminate against out-of-staters." If applied more broadly, the court said, the doctrine might end up penalizing states for statutes that either did not affect or actually benefitted interstate commerce:

> [A]s far as we know, all fossil fuel producers in the area served by the grid will be hurt equally and all renewable energy producers in the area will be helped equally. If there's any disproportionate adverse effect felt by out-of-state producers or any disproportionate advantage enjoyed by in-state producers, it hasn't been explained to this court. And it's far from clear how the mandate might hurt out-of-state consumers either. . . . To reach hastily for *Baldwin*'s *per se* rule, then, might lead to the decidedly awkward result of striking down as an improper burden on interstate commerce a law that may not disadvantage out-of-state businesses and that may actually reduce price for out-of-state consumers.[24]

Notably, the opinion was written by Neil Gorsuch, who was later appointed by President Trump to the Supreme Court.

The concept of extraterritoriality is certainly less than crystal clear, but it would be a mistake to apply it expansively. In a unified national market, any important state regulation is likely to have some spillover effects on other markets. The extraterritoriality prong of dormant commerce clause doctrine is strong medicine. Courts have employed it sparingly and only in cases where the state exercised effective control over transactions wholly outside its borders. Many state regulations have *some* impact on producers or consumers outside the state; such impacts are the very reason for the balancing test. To the extent that the extraterritoriality doctrine is aimed at

24 Energy & Env't Legal Inst. v. Epel, 793 F.3d 1169 (10th Cir. 2015).

those impacts, it is a crude tool compared with the *Pike* balancing test. If the extra-territoriality doctrine is given its proper, narrow scope, it should not pose a challenge to well-designed state efforts to regulate embedded carbon in locally consumed goods. But extraterritoriality may be to some extent in the eye of the beholder, and judges may think extraterritoriality is present unless they understand the reasons for regulating embedded carbon.

We may soon get some guidance from the Supreme Court on the scope of this doctrine. In May of 2022, the Court agreed to hear another Ninth Circuit case, this one involving a restriction on the sale of pork from pigs that were not humanely raised.[25] The Ninth Circuit rejected an extraterritoriality challenge to the law. The Court's opinion in the case may clarify the boundary between laws that violate the extraterritoriality ban and those that merely have effects on out-of-state producers.

III. Other Federalism Issues

Although the dormant commerce clause is the potential challenge with the broadest applicability, state climate regulations may face other challenges. There is extensive federal regulation of energy markets, and it is not hard for industry to formulate arguments that state energy regulations conflict with these federal schemes. In addition, state actions may be subject to challenges under the Compact Clause when the state(s) attempt to cooperate with other jurisdictions, as well as to challenges under the foreign affairs preemption doctrine when the other jurisdiction is foreign. All of this provides a fertile area for litigation.

A. Statutory Preemption

Part II dealt with the validity of state regulation independent of federal regulation. In this section we will be concerned with the validity of state regulations in areas where Congress has acted. In cases of direct conflict, the state statute must give way. The Supremacy Clause of the Constitution provides:

> This Constitution, and the Laws of the United States which shall be made in Pursuance thereof; and all Treaties made, or which shall be made, under the authority of the United States, shall be the Supreme Law of the Land; and the Judges in every State shall be bound thereby, any Thing in the Constitution or Laws of any State to the Contrary notwithstanding.

[25] Nat'l Pork Producers Council v. Ross, 6 F.4th 1021 (9th Cir. 2021), cert. granted, 142 S. Ct. 1413 (2022).

Thus, a federal statute trumps state law, assuming the federal statute is constitutional.

Congress often speaks directly to the scope of state authority to regulate. Congress can limit state authority explicitly, for instance by stating that any state law or administrative action in a particular area is invalid. Conversely, Congress can preserve state laws through a savings clause, such as a statement that nothing in a particular statute shall preclude states from taking particular types of action. But courts can find that a state law is preempted despite the lack of explicit preemption language in a statute. In the climate change arena, two doctrines concerning "implied preemption" are most relevant.

First, courts may rule that an entire domain is off-limits to states, which is often called field preemption. One basis for field preemption is that the federal regulatory scheme may be so pervasive and detailed as to suggest that Congress left no room for states to supplement it. Alternatively, a statute enacted by Congress may involve a field in which the federal interest is so dominant that enforcement of all state laws is precluded. Other aspects of the regulatory scheme imposed by Congress may also support the inference that Congress has completely foreclosed state legislation in a particular area, including even state laws that support federal goals. To date, the courts have not implied field preemption from environmental statutes, but state energy regulations could run up against field preemption under some non-environmental federal statute such as the Federal Power Act's ban on state regulation of wholesale electricity transactions.

Second, even when Congress has not completely foreclosed state regulation on a subject, a state statute is void to the extent that it actually conflicts with a valid federal statute. Such a conflict can be found where compliance with both the federal and state regulations is impossible, or more often, where the state law interferes with Congress's objectives. Conflict preemption often involves difficult judgments about congressional purposes and the permissible degree of incidental interference by state laws.

Preemption cases are first and foremost about statutory interpretation. If there is an express preemption clause or a savings clause, the court must determine the scope of the clause. This can be a difficult undertaking. When the statute is not explicit, the court must make a difficult judgment of its own, in effect deciding whether Congress would have wanted the state law preempted if it had anticipated the situation. Not surprisingly, preemption decisions are often difficult to parse and are a fertile source of disagreement among judges. There are also ideological cross-currents, because the judges perspectives can be shaped by their views of federalism and of government regulation, which may point in opposite directions in these cases.

Because the Clean Air Act is the basic federal statute regulating emissions into the air (including greenhouse gases), it is the obvious starting point in thinking about possible preemption of state climate regulations. The Clean Air Act, like almost all contemporary federal environmental laws, utilizes a basic "floor" preemption strategy. Federal law sets minimum required levels of environmental protection (the "floor"), but the states are expressly authorized to go further and adopt more stringent environmental requirements. Like many other statutes, the Clean Air Act contains an explicit savings clause to limit preemption. The language of the Clean Air Act's savings clause is quite sweeping. Section 116 provides that "nothing in the chapter shall preclude or deny the right of any State or political subdivision thereof to adopt or enforce (1) any standard or limitation respecting emissions of air pollutants or (2) any requirement respecting control or abatement of air pollution."[26] Thus, states seem to be free to impose stricter limits on carbon emissions than the federal government.

There is an important exception to the savings clause in section 209 of the statute, which applies to regulations of emissions from new vehicles. The Clean Air Act directs EPA to issue federal standards for tailpipe emissions from vehicles. When the statute was under consideration, the automobile industry was alarmed at the risk that it would have to produce multiple models of cars to meet emissions standards in different states. Section 209 responds to that concern. Subsection (a) of section 209 prohibits states and their subdivisions from adopting or enforcing any standard relating to emissions controls from new vehicles. Taken alone, section 209(a) would seem to completely preempt state regulation, a classic case of field preemption. But section 209(b) creates an important exception to this preemption rule. Although it does not mention California by name, section 209(b) is drafted in a way that allows only California to qualify. It permits California to apply for a waiver in order to adopt stricter standards based on "compelling and extraordinary" circumstances. The rationale was that Southern California's severe air pollution problems were likely to be insoluble unless the state was allowed to vigorously regulate pollution from cars and trucks.

By itself, section 209(b) would allow only California to impose such regulations. But given that manufacturers would have to set up production runs to supply the California market, it was relatively feasible for them to supply similar vehicles to other states. For that

---

[26] Clean Air Act § 116, 42 U.S.C. § 7416.

reason, Congress later decided to allow other states to piggyback on the California standards. Under section 177, other states have the option of adopting standard's identical to California's, with no deviations allowed.[27] The upshot is that car manufacturers produce a "national car" complying only with the federal standards and a "California car" meeting that state's higher standards.

So far, the most important preemption dispute relating to climate change under the Clean Air Act involved California's efforts to impose tailpipe standards for greenhouse gases. The Bush Administration denied a waiver, but the Obama Administration reversed course. Car manufacturers immediately challenged the validity of the waiver, but the litigation was settled as part of an agreement with the manufacturers over federal tailpipe standards. The Trump Administration withdrew the Obama waiver. Several major companies entered into agreements with California to continue to follow California's standards regardless, leaving the industry divided. In March of 2022, the Biden EPA reinstated the waiver.

It is important to keep in mind that section 209 does not give California carte blanch to regulate new vehicles. First, the statute applies to statewide standards only, not regulations by particular air quality districts or municipalities. Second, California must actually apply for a waiver—it is not enough that it could have obtained a waiver if it asked. And third, California must actually obtain the waiver. Similarly, other states can piggyback on California's waiver but they must be careful to follow California's lead precisely, since deviations can be fatal to their authority under section 177.

Note that this preemption provision applies only to the transportation sector. The Clean Air Act is unlikely to be a source of preemption in the electricity sector, given the savings clause. However, another statute, the Federal Power Act (FPA), can be a potential issue. The FPA gives the Federal Energy Regulatory Commission (FERC) jurisdiction over interstate transmission and over wholesale interstate transactions. The courts have interpreted FERC's authority over interstate wholesale transactions to be exclusive, in effect applying field preemption. But the FPA also has a strong savings clause. Section 201 provides that FERC "shall not have jurisdiction . . . over facilities used for the generation of electric energy or over facilities used in local distribution or only for the transmission of electric energy in intrastate commerce, or over facilities for the transmission of electric energy consumed wholly by the transmitter." The clear intention is to draw a line between local utility regulation in the retail market and federal regulation in the

---

[27] Clean Air Act § 177, 42 U.S.C. § 7507.

interstate wholesale market. The trouble is that this line between retail and wholesale is not always terribly clear.

It has become even harder to draw that line as federal regulation of the interstate electricity market has moved away from traditional utility regulation, where the regulatory agency sets a reasonable price for electricity or electricity transmission. FERC has basically abandoned that approach in favor of trying to structure electricity markets to produce competitive prices. State regulations can impact market structure or operations even if they do not directly interfere with price. This makes it harder to distinguish between the state's legitimate interests in generation and consumer distribution within its boundaries and the FERC's control of the interstate market. For that reason, FPA preemption claims are endemic in cases involving challenges to state renewable energy laws. A more detailed discussion of these issues can be found in Chapter 5.

B. Foreign Affairs Preemption

In statutory preemption cases, at least courts have the advantage of a specific statutory scheme to guide the inquiry. A knottier issue arises in the context of state agreements with foreign jurisdictions: whether such agreements infringe the exclusive federal power over foreign affairs.[28] Since climate change is a global issue, it is not surprising that states have made connections with foreign countries in attempting to deal with the issue.

Even apart from any specifically international activity on the part of states, the very fact that state governments are addressing a global issue may seem incongruous. We generally expect states to deal with local problems, not to address the problems of the world. Even people who are unfamiliar with constitutional law probably realize that there is an American embassy in foreign countries, not separate embassies for each state, and that by the same token it is the President and Secretary of State who negotiate treaties, not their state counterparts. The issue, then, is whether state efforts at international cooperation cross over into the exclusively federal domain of foreign affairs.

It is helpful to start with a brief review of the constitutional provisions relating to international affairs. The Constitution gives various organs of the federal government the authority to enter into treaties, receive ambassadors, and go to war. Other provisions ban the states from making war or entering treaties (but not necessarily

[28] *See* Note, *Foreign Affairs Preemption and State Regulation of Greenhouse Gas Emissions*, 119 HARV. L. REV. 1877 (2006). A concise doctrinal overview can be found in Judith Resnick, *Foreign as Domestic Affairs: Rethinking Horizontal Federalism and Foreign Affairs Preemption in Light of Translocal Internationalism*, 57 EMORY L.J. 31, 71–78 (2007).

"agreements" or "compacts" with foreign states). The upshot is to give the federal government control over foreign affairs. Nevertheless, in an increasingly globalized society, states cannot completely ignore the world outside the United States. It seems to be increasingly common for states to reach out beyond national borders in their activities, and not just in the area of climate change. For instance, state governors often lead trade delegation to foreign countries in the hope of encouraging foreign investment within their states. Recent Supreme Court cases leave great doubt, however, about the limits on states' ability to engage with foreign jurisdictions.

In the past two decades, the Supreme Court has issued several opinions dealing directly with implied restrictions on state regulatory authority affecting foreign affairs.[29] None of the cases involved environmental issues, but they provide the legal parameters that apply to transnational activities by states.

The first case was *Crosby v. National Foreign Trade Council*.[30] A Massachusetts law prohibited state or local governments from doing business with companies that were themselves doing business with Burma (now Myanmar). The Court concluded that the state law interfered with a provision of the federal law that gave the President discretion to control economic sanctions against Burma. Also, the state sanctions were harsher than the federal sanctions, trespassing beyond the limits Congress had set for pressuring Burma. As the Court said, the state laws "compromise the very capacity of the President to speak for the Nation with one voice in dealing with other governments."[31]

The Court's later ruling in *American Insurance Ass'n v. Garamendi*,[32] is more difficult to interpret. The case deals with the repercussions of the Holocaust. Many World War II era insurance policies held by European Jews were either confiscated by the Nazis or dishonored by insurers after the war. Ultimately, the Allied governments had mandated restitution to Nazi victims by the (then) West German government. Unfortunately, although a large numbers of claims were paid, many others were not, and large-scale litigation resulted after German reunification. Unlike *Crosby*, there was no federal statute bearing on the issue. However, the U.S. had entered into negotiations to try to resolve the dispute, resulting in an agreement with Germany. In the meantime, California passed a law

---

[29] An earlier significant decision was Zschernig v. Miller, 389 U.S. 429 (1968), which struck down an Oregon law that allowed aliens to inherit property in that state only if their home country allowed Americans to inherit the property of that country's citizens.

[30] 530 U.S. 363 (2000).

[31] 530 U.S. at 381.

[32] *Id.* at 381.

requiring any insurer doing business in the state to disclose information about all policies sold in Europe between 1920 and 1945.

A narrowly divided Court struck down the California law.[33] According to the majority, the consistent presidential policy had been to encourage voluntary settlement funds in preference to litigation or coercive sanctions. California sought to place more pressure on foreign companies than the president had been willing to exert. This clear conflict between the federal policy and state law was itself a sufficient basis for preemption. The majority found the argument for the preemption issue particularly persuasive given the weakness of the state's interest in terms of traditional state legislative activities.[34] Some of the language in the opinion could suggest a broad interpretation of foreign affairs preemption.

Garamendi's broader language about Presidential authority to preempt state law may have been tempered by the later decision in *Medellin v. Texas*.[35] *Medellin* was a complex case involving state violation of an international consular treaty, as interpreted by a decision of the International Court of Justice (ICJ). The Supreme Court found that the treaty providing the ICJ's authority to decide cases involving the United States did not bind the states directly. But President George W. Bush had also issued a memorandum stating that the "United States will discharge its international obligations . . . by having State courts give effect to the [ICJ] decision in accordance with general principles of comity." The Court rejected the President's argument that he had inherent authority "to establish binding rules of decision that preempt state law." The scope of *Garamendi* was not directly at issue, but the Court was clearly less sympathetic to the argument that Presidential actions had the effect of preempting state law without any support from a treaty, statute, or clear history of Congressional approval.

It seems unlikely that courts will apply foreign affairs preemption to rule out state laws that independently limit carbon emissions. But states do not always attempt to reduce carbon emissions on their own. Instead, they seek to work with other state

[33] The majority included a conservative (Chief Justice Rehnquist) and four of the Court's centrist judges (Souter, O'Connor, Kennedy, and Breyer), while the dissent contained two liberals (Ginsburg and Stevens) as well as the Court's two most conservative members (Scalia and Thomas).

[34] The dissent cogently argued that upholding the state law "would not compromise the President's ability to speak with one voice for the Nation" and that the Court should reserve foreign affairs preemption for "circumstances where the President, acting under statutory or constitutional authority, has spoken clearly to the issue at hand." *Id.* at 442.

[35] 552 U.S. 491 (2008).

governments or even with foreign jurisdictions, as exemplified by the Quebec-California emissions trading system.

When states try to pursue their policy goals through cooperation with foreign governments, they may encounter additional constitutional problems. Recall that states are forbidden to enter into treaties with foreign governments. In addition, the Compact Clause provides that no state can enter into an "Agreement or Compact with another State, or with a foreign Power" without the approval of Congress. As we will see in the next section, however, the term *Agreement* does not include all cooperative arrangements. The Supreme Court cases involve cooperation between states rather than cooperation with foreign jurisdictions. But given that the same constitutional language applies in both the interstate and transnational settings, the doctrine should apply equally to agreements between state governments and foreign jurisdictions.

C. The Compact Clause

Just as states have made linkages with foreign jurisdictions in their climate mitigation efforts, they have also made linkages with each other. State emissions trading systems often involve linkages with other states or Canadian provinces. For instance, in 2001, a group of New England Governors and Eastern Canadian Premiers agreed to reduce greenhouse gas emissions to 1990 levels by 2010 and 10 percent below those levels by 2020.[36] Such linkages offer several advantages. States can "magnify the importance of their climate change initiatives by banding together with other state and local governments."[37] In an emissions trading system, "size matters" because "a greater number of sources makes possible a greater number of trades thus making the market more competitive."[38] Finally, regional systems may fit well with the regional organization of the electricity grid.[39]

Despite their advantages, regional agreements may encounter constitutional challenges. The main issue is whether congressional consent to a regional agreement is required under the Compact Clause. To what extent does this clause prevent cooperation by states with other states or foreign jurisdictions—for example, in jointly designing and implementing emission trading systems? The answer seems to be the Compact Clause should not be a major problem for states pursuing linkages with other jurisdictions.

---

36 Engel, *supra* note 18, at 65.

37 *Id.* at 64.

38 *Id.* at 69.

39 *Id.* at 71.

The Supreme Court has not construed the Compact Clause to reach all agreements between states, but only those that are "directed to the formation of any combination tending to the increase of political power in the States, which may encroach upon or interfere with the just supremacy of the United States."[40] On this basis, the Court upheld the formation of a multistate tax commission formed to develop tax policies for individual states, which would then be adopted separately by each member state.[41] The commission had the power to conduct audits using subpoenas in any of the member states' courts, including audits of multinational corporations.

Similarly, in *Northeast Bancorp, Inc. v. Bd. of Governors of the Federal Res. Sys.*,[42] the Court found that no compact existed despite states' adoption of identical laws and informal agreements between state officers regarding acquisition of local banks by out-of-state banks. Although the parallel state laws were adopted in concert, the Court found other circumstances more important: that no joint regulatory body was established, the statutes were not conditional on each other, and states were not legally bound.[43] The Court held that the statutes did not "either enhance the political power of the New England States at the expense of other States or have an 'impact on our federal structure.' "[44] Note that the text of the Compact Clause does not distinguish between other states and other countries, so the reasoning of *Northeast Bancorp* would appear to apply in both contexts.

In designing trading systems, states have been careful to follow these guidelines. The Northeast trading system, RGGI, was the product of two years of negotiations between states. The governors of the states entered into a Memorandum of Understanding (MOU), which ultimately led to the creation of a model rule for adoption by individual states. States then individually adopted regulations based on the model rule. At no point were the states as sovereign entities legally bound to take any action, nor did they delegate regulatory power to an interstate entity. In fact, states have moved in and out of the agreement depending on local politics. All of this is in line with the Supreme Court's rulings upholding the multi-state tax commission and bank acquisition agreements.

A virtually forgotten provision of the Clean Air Act should soothe any remaining doubts about the legality of interstate linkages such as RGGI, although it does not apply to linkages with foreign

---

40 Virginia v. Tennessee, 148 U.S. 503, 519 (1893).

41 United States Steel Corp. v. Multistate Tax Comm'n, 434 U.S. 452 (1978).

42 472 U.S. 159 (1985).

43 *Id.* at 175.

44 *Id.* at 176.

jurisdictions. Section 102 of the Clean Air Act is entitled "cooperative activities."[45] Subsection (a) calls upon EPA to encourage "cooperative activities by the States and local government" and foster the passage of uniform state laws. Subsection (c) is even more clearly on point. It provides:

> The consent of the Congress is hereby given to two or more States to negotiate and enter into agreements or compacts . . . for (1) cooperative effort and mutual assistance for the prevention and control of air pollution and the enforcement of their respective laws relating thereto, and (2) the establishment of such agencies, joint or otherwise, as they may deem desirable for making effective such agreements or compacts. No such agreement or compact shall be binding or obligatory upon any State a party thereto unless and until it has been approved by Congress.[46]

This provision squarely covers RGGI-like interstate agreements, given that the Supreme Court has held that greenhouse gases are a form of air pollution under the statute. Congressional consent is needed only in order to make an agreement about greenhouse gases legally binding on the states. Thus, although states may retain the right to withdraw, an interstate trading agreement seems permissible even if it goes beyond the safe harbor provided by the Supreme Court opinions.[47]

Because larger emissions trading systems are more effective, it is advantageous to link different systems together. For that reason, California has linked its emission trading system with Quebec's. The Trump Administration filed suit to invalidate this linkage. The district court ruled in favor of California, holding that the linkage agreement did not violate the Compact Clause or the Treaty Clause.[48] It did not involve sufficiently weighty matters to constitute a treaty, and was not a compact given California's unilateral right to withdraw. In a separate opinion, the court also held that foreign affairs preemption did not apply because California's linkage was not

---

[45] CAA § 102, 42 U.S.C. § 7402.

[46] CAA § 102(c), 42 U.S.C. § 7402(c). A concluding sentence provides that compacts relating to "control and abatement of air pollution in any air quality control region" can only include states in that region. That sentence seems to have no application to climate change, which does not relate to a specific air quality control region.

[47] Section 102 may also be relevant to certain kinds of discrimination claims. States outside the agreement (and their firms) can hardly complain that they fail to receive the benefits of an agreement that they have not entered.

[48] United States v. California, 444 F. Supp. 3d 1181 (E.D. Cal. 2020).

primarily designed to control conduct outside its borders and did not directly conflict with U.S. foreign policy.[49]

IV. Assessing State Mitigation Actions

How much can state policies contribute to reducing greenhouse gases? Clearly, the extent that any state can contribute towards solving such a global problem is limited. Even California, which has one of the ten largest economies in the world, is a minor emitter compared with the United States as a whole or China or the European Union. Cities, of course, are even smaller players. Still, even small reductions in greenhouse gases have some positive effect: the less carbon in the atmosphere, the lower the amount of harm from climate change.

The ability of states to reduce pollution is limited, however, by the problem of leakage, which was discussed in Chapter 2. For instance, if a state were to close its own coal-fired power plants, it might find that utilities simply purchased electricity on the wholesale market from coal-fired plants in neighboring states. Similarly, if the state imposes restrictions on manufacturers in its own state, manufacturing might simply increase in another state that does not impose such emissions standards. Because states are relatively small and because goods, services, and capital move readily between states, leakage is a larger problem at the state level than at the national level. It is difficult to estimate the magnitude of leakage, but it is clearly not insignificant.

States can take some steps to reduce leakage. They can impose less stringent restrictions on industries where leakage seems more likely to be a problem. They may also have some ability to reduce leakage by regulating the carbon intensity of imports, as shown in the *Rocky Mountain* case. But some leakage is almost inevitable given that states operate in a wide-open national market. This leakage needs to be offset against the emission reductions achieved within the state.

Leakage means that regulation in one states increases emissions elsewhere. But the opposite effect is also possible. Localities do not act entirely on their own when they choose to address climate change. Both within the United States and globally, there are well-developed networks of countries, state governments, and cities that provide mutual support and coordination. Moreover, like early adopters of new technologies, these "early adopters" of climate policy can promote improvements in regulatory tools, which make it easier for others to adopt those tools later. More directly, by

[49] United States v. California, 2020 WL 4043034, at *2 (E.D. Cal. July 17, 2020), appeal dismissed, No. 20-16789, 2021 WL 4240403 (9th Cir. Apr. 22, 2021).

providing markets for clean energy technologies, they promote the development of those technologies and help producers gain the experience necessary to reduce costs through "learning by doing." In these ways, early adopters may make it more likely that others will follow in their footsteps.

Some scholars argue that this "bottom up" approach to climate mitigation has better prospects than the top-down approach in which an international agreement sets standards, which are then adopted at the national level. The problem with the top-down approach is that it requires universal agreement, or at the very least agreement of all the largest emitters such as the United States, the EU, China, India, and perhaps another half dozen major countries. That agreement has proved difficult to establish and maintain in practice. Advocates of the bottom-up approach argue that it may be easier to create a global climate regime by building cooperative networks of jurisdictions. These networks could gradually grow until they controlled a sufficient share of emissions to bring the problem under control.

As a practical matter, subnational governments like states and cities have often been in the forefront of climate policy, and their efforts seem to be gathering momentum. At one time, California was an outlier in its support for aggressive climate action. Now, even though California remains a leader, many other states have set ambitious targets, and some states are ahead of California on some issues. Elsewhere in the world, states or provinces along with cities have also taken leadership roles. At this point, it seems clear that climate policy is by no means just the business of national governments. If anything, subnational jurisdictions are forging the path for national governments to follow behind.

State and local governments also have another important role to play in the climate change arena. Up to this point in the book, we have focused on efforts at all levels (international, national, state) to mitigate climate change by reducing carbon emissions. But as we saw in Chapter 1, some degree of climate change seems virtually inevitable. Society will then be faced with the problem of adapting to these changes. Because many of the impacts of climate change are localized, much of the adaptation effort is likely to fall on state and local governments. Thus, the next chapter, which deals with climate change adaptation, involves matters that are directly relevant to these levels of government.

Further Readings

Brannon P. Denning, *Environmental Federalism and State Renewable Portfolio Standards*, 44 CASE W. RES. L. REV. 1519 (2014).

KIRSTEN ENGEL, EDELLA C. SCHLAGER, & SALLY RIDER, NAVIGATING CLIMATE CHANGE POLICY: THE OPPORTUNITIES OF FEDERALISM (2011).

Daniel A. Farber, *Carbon Leakage Versus Policy Diffusion: The Perils and Promise of Subglobal Climate Action*, 13 CHI. J. INT'L LAW 359 (2013).

Daniel A. Farber, Yuichiro Tsuji, and Shiyuan Jing, *Thinking Globally, Acting Locally: Lessons from the U.S., Japan, and China*, 93 OHIO ST. L.J. 953 (2021).

Alexandra B. Klass and Elizabeth Henly, *Energy Policy, Extraterritoriality, and the Dormant Commerce Clauses*, 5 SAN DIEGO J. CLIMATE & ENERGY L. 69 (2013–2014).

Keith H. Hirokawa, Jonathan Rosenbloom & Michelle Zaludek, *Climate Change Adaptation and Land Use Law*, in RESEARCH HANDBOOK ON CLIMATE CHANGE ADAPTATION LAW, Jonathan Verschuuren ed. (2022).

Felix Mormann, *Constitutional Challenges and Regulatory Opportunities for State Climate Policy Innovation*, 41 HARV. ENV'T L. Rev. 189 (2017).

Hari Osofsky, *The Geography of Solving Global Environmental Problems: Reflections on Polycentric efforts to Address Climate Change*, 58 NYL SCH. L. REV. 777 (2013/2014).

Eleanor Stein, *Regional Initiatives to Reduce Greenhouse Gas Emissions*, in MICHAEL B. GERARD AND JODY FREEMAN (EDS.), GLOBAL CLIMATE CHANGE AND U.S. LAW (2d ed. 2014).

Chapter 8

CLIMATE IMPACTS AND ADAPTATION

This chapter will examine the concept of adaptation and the efforts taken to advance adaptation goals at the domestic and international levels. We will begin by reviewing the types of challenges that climate change poses, how adaptation has emerged as a frame for thinking about ways to prepare for and respond to these challenges, and how adaptation efforts have evolved as the processes of climate change have progressed.

I. Adaptation Needs

During the first three decades of diplomacy, global efforts to address climate change focused primarily on mitigation, or those efforts designed to reduce anthropogenic forcing of the climate system, primarily through efforts to control greenhouse gas emissions. Mitigation efforts were the focus of attention for so long for a number of reasons. During the first decade of global negotiations (1990–2000), the focus was on understanding the patterns of global climate change and curbing its causes. Instigated in the wake of the adoption of the Montreal Protocol on Substances that Deplete the Ozone Layer—which would go on to become one of the most effective multilateral efforts to address a global environmental problem—early climate change negotiations took place when curbing climate change seemed politically and scientifically feasible. As the extent of the technological, economic, and political challenge became evident, mitigation continued to receive the brunt of political attention for a variety of reasons, including concerns that focusing on adaptation would detract from mitigation efforts, signal defeat, create complex questions about who should pay for adaptation efforts, require difficult decisions to be made about where to allocate resources, and even displace or undermine existing forms of development aid.[1] In essence, adaptation was viewed as complicated, expensive, and in tension with mitigation. For all of these reasons, while efforts to facilitate adaptation are now the focus of a significant amount of attention, this has not always been the case.

Before continuing, it is helpful to understand what adaptation efforts encompass. The IPCC defines adaptation as an "adjustment

---

[1] *See, e.g.*, Michael B. Gerrard, Introduction and Overview, in THE LAW OF ADAPTATION TO CLIMATE CHANGE: U.S. AND INTERNATIONAL ASPECTS, Michael B. & Katrina Fischer Kuh eds., pp 3–4 (2012).

in natural or human systems in response to actual or expected climatic stimuli or their effects, which moderates harm or exploits beneficial opportunities."[2] In the legal literature, climate change adaptation has been similarly characterized as those efforts "designed to increase the resilience of natural and human ecosystems to the threats posed by a changing environment."[3] Adaptation is used broadly to refer to all of the varied efforts undertaken at multiple levels of governance to prepare for climate change, whether the intent is to maximize opportunities or minimize risks associated with climate change. Moreover, as JB Ruhl describes, there are two primary forms of climate impacts driving adaptation responses: "changes in the extremes, frequency, and distribution of environmental attributes such as temperature, precipitation, and wildfire" that erode 'stationarity' in our traditional management time-frames, and "new challenges for which we have no prior history of adaptation" such as sea level rise and mass climate-induced migration.[4] Ruhl refers to this second category as "the unprecedented; 'no-analog' phenomena of our climate change future."

The gradual shift away from mitigation-oriented negotiations towards a more balanced focus on adaptation has been spurred by failed efforts to develop successful mitigation strategies and increased data about the pace and inevitability of some degree of warming. In the Sixth Assessment Report (AR6), the IPCC confirmed that "[i]t is unequivocal that human influence has warmed the atmosphere, ocean and land" and that human-induced climate change is "affecting many weather and climate extremes in every region across the globe" bringing about "changes in extremes such as heatwaves, heavy precipitation, droughts, and tropical cyclones."[5] Projected climatic changes create a series of risks that will necessitate adaptation efforts, including: temperature increases, extreme weather events, sea level changes, drought, food scarcity, fresh water shortages, extreme precipitation, changing patterns of disease, ocean acidification, and wildfires. Each of these areas of risk poses direct and indirect threats to human and natural systems.

---

[2] INTERGOVERNMENTAL PANEL ON CLIMATE CHANGE, CLIMATE CHANGE 2007: IMPACTS, ADAPTATION AND VULNERABILITY 6 (M.L. Parry et al. eds., 2007), https://www.ipcc.ch/site/assets/uploads/2018/03/ar4_wg2_full_report.pdf.

[3] Robert L. Glicksman, *Climate Change Adaptation: A Collective Action Perspective on Federalism Considerations*, 40 ENVTL. L. 1159, 1159 (2010).

[4] J.B. Ruhl, *Climate Adaptation Law*, in GLOBAL CLIMATE CHANGE AND U.S. LAW (Michael B. Gerrard & Jody Freeman eds., 3rd ed, forthcoming 2022) [hereinafter Ruhl, *Climate Adaptation Law*].

[5] INTERGOVERNMENTAL PANEL ON CLIMATE CHANGE, CLIMATE CHANGE 2022: THE PHYSICAL SCIENCE BASIS, HEADLINE STATEMENTS 1 (2022), https://www.ipcc.ch/report/ar6/wg1/downloads/report/IPCC_AR6_WGI_Headline_Statements.pdf [hereinafter HEADLINE STATEMENTS].

While it is not possible to discuss in detail all of the different threats that climate change poses, a few details about two separate but interrelated areas of change help contextualize the risks climate change poses to human and natural systems. With respect to food security, for example, projected patterns of climate change are expected to threaten fisheries and impair growing conditions for some of the most important global food staples, including wheat, rice, and maize. As a result of increased pressures on marine and terrestrial food systems, coupled with increasing food demands associated with population growth patterns, global food security could be severely undermined by the middle of the century. Similarly, climate change is also expected to lead to reductions in surface and groundwater resources, reducing the total amount of water available for key uses such as drinking water, irrigation, and energy production. Water shortages will compound food security challenges and shortages in both areas will intensify competition for scarce resources with expected implications for domestic and international security.

Alongside the direct threats that climate change poses, including heat stress, food scarcity, water scarcity, extreme precipitation events, there are growing concerns about the indirect effects of climate change. Indirect effects might include growing incidences of mental health challenges associated with lost livelihoods or economic stress, increased patterns of violence, including gender-based violence associated with physical and economic stress, and unparalleled rates of physical displacement. With respect to this latter concern, it is projected that climate change will increase patterns of migration and displacement of people both within and across national borders creating a myriad of social, economic, and legal challenges. For example, the most recent IPCC report confirmed that climate change "is contributing to humanitarian crises" and that "[c]limate and weather extremes are increasingly driving displacement,"[6] with displacement and involuntary migration likely to increase over time.[7]

While this Chapter and much of the literature focuses on the threats that climate change poses to human systems, it is also important to highlight the threats that climate change poses to global biodiversity. The IPCC's Sixth Assessment Report highlights the intensifying threat to global biodiversity, warning that "[c]limate-caused local population extinctions have been widespread among

---

[6] HANS-O. PÖRTNER ET AL., INTERGOVERNMENTAL PANEL ON CLIMATE CHANGE, CLIMATE CHANGE 2022: IMPACTS, ADAPTATION, AND VULNERABILITY: SUMMARY FOR POLICYMAKERS B.1.7 (2022), https://report.ipcc.ch/ar6wg2/pdf/IPCC_AR6_WGII_SummaryForPolicymakers.pdf [hereinafter Summary for Policymakers].

[7] *Id.* at B.4.7.

plants and animals" and that "at warming levels beyond 2°C by 2100, risks of extirpation, extinction and ecosystem collapse escalate rapidly."[8] The threats to global biodiversity are significant in their own right while also having implications for human health and well-being.

As the IPCC summarizes it in the Sixth Assessment Report, "the scale of recent changes across the climate system as a whole—and the present state of many aspects of the climate system—are unprecedented over many centuries to many thousands of years."[9] At this point, however, it is important to highlight that the nature and extent of the risks that climate change poses to systems and communities differs from place to place and depends on a variety of factors. Increasingly, we understand the total risk that a community faces as resulting from an interaction of factors, most prominently the interaction of climate-related hazards, vulnerability, and exposure.[10] Climate related hazards include threats ranging from sea level rise, storm surge, drought, and changing patterns of disease. Understanding and predicting the pathways of these hazards is only the first step to understanding what types of risk the hazard poses for a community. Next, it is necessary to understand the extent to which a community is vulnerable to the hazard, or the degree to which the community is predisposed to be negatively affected based on its "sensitivity or susceptibility to harm and lack of capacity to cope and adapt."[11] The extent to which a community is vulnerable depends on a number of factors, including political stability, economic stability, disaster response systems, and community cohesion. Finally, it is necessary to assess the level of exposure the community faces, or the degree to which natural and human capital rest in the pathway of harm. Only by assessing the interaction between hazard, vulnerability, and exposure is it possible to begin to calculate the risk that communities face with respect to individual and cumulative climate-related changes.

The site- and impact- specific nature of risk assessment renders adaptation planning immensely complex, particularly in areas where communities face multiple and interacting hazards and high levels of both vulnerability and exposure, which is the case for many developing and least developed countries in the Global South. As one commentator notes, "[c]atastrophe is bad for everyone. But it is

---

[8] INTERGOVERNMENTAL PANEL ON CLIMATE CHANGE, WORKING GROUP II—IMPACTS, ADAPTATION AND VULNERABILITY: FACT SHEET—BIODIVERSITY, https://report.ipcc.ch/ar6wg2/pdf/IPCC_AR6_WGII_FactSheet_Biodiversity.pdf.

[9] HEADLINE STATEMENTS, *supra* note 5.

[10] *Id.* at 3.

[11] *Id.* at 5.

especially bad for the weak and disenfranchised."[12] This is particularly true with respect to climate change, making the need for adaptation planning that much more pressing. As one discrete example, for a country such as Bangladesh, which is considered to be at high risk absent adaptation efforts, climate change could prove devastating. Bangladesh is a South Asian country that is classified by the United Nations as a Least Developed Country, due to its low gross national income, weak human capital, and high degree of economic vulnerability. Bangladesh also has high levels of exposure to hazards such as sea level rise and extreme events since a large percentage (upwards of 50%) of its population lives within 10 miles of the coastline, with estimates suggesting that upwards of 27 million people could be living along the exposed coastline by 2050. As a result, Bangladesh "is identified as being at specific risk from climate change due to its exposure to sea-level rise and extreme events and concentrated multidimensional poverty," with discreet risks included those posed by: cyclones, food insecurity, heat stress, increased incidence of disease, and water stress as a result of salt water intrusion, to name a few.[13] Given the inevitability of continuing sea-level rise, increased frequency and intensity of storms, and increasing temperatures, the long-term wellbeing of millions of citizens in Bangladesh may turn on the ability of communities to adapt to increasing levels of risk.

Moving from the global to the domestic level, the US National Climate Assessment frames the urgency and the widespread nature of the challenge, noting that:

> [The] impacts of climate change are already being felt in communities across the country. More frequent and intense extreme weather and climate-related events, as well as changes in average climate conditions, are expected to continue to damage infrastructure, ecosystems, and social systems that provide essential benefits to communities. Future climate change is expected to further disrupt many areas of life, exacerbating existing challenges to prosperity posed by aging and deteriorating infrastructure, stressed ecosystems, and economic inequality.[14]

---

[12] ROBERT R.M. VERCHICK, FACING CATASTROPHE: ENVIRONMENTAL ACTION FOR A POST-KATRINA WORLD 104 (2010).

[13] International Center for Climate Change and Development, *Briefing: What Does the IPCC Say About Bangladesh?* (Oct. 2014), 2–3, http://www.icccad.net/wp-content/uploads/2015/01/IPCC-Briefing-for-Bangladesh.pdf.

[14] U.S. Global Change Research Program, *Fourth National Climate Assessment: Volume II: Impacts, Risks, and Adaptation in the United States*, Summary Findings (2018), https://nca2018.globalchange.gov.

Across the United States, not only are our summers growing longer and hotter and our winters shorter and warmer, but also rain events are becoming more extreme, seasonal allergies are lasting longer and becoming more intense and a host of other effects are equally severe. In coastal areas, residents are experiencing more regular and more extreme flooding. So, too, are residents of inland cities near large rivers. While some parts of the United States experience intensifying storm and flooding events, many areas of the Western United States are becoming hotter and drier and, thus, more susceptible to wildfires that are increasing in frequency and intensity. Beyond the mainland, in Arctic Alaska, sea ice is receding with subsequent effects on wildlife and storm patterns, all of which combines to place severe pressures on many Arctic communities, with some communities now being forced to relocate in order to survive. In Hawaii, already constrained freshwater supplies are facing additional stressors and increased temperatures are amplifying pressures on human and natural systems and exacerbating concerns around both food and water security.

The effects of climate change reach far and wide and pose a complex set of risks for natural and human systems in both the short and long-term. Growing understanding of this reality has prompted policymakers at all levels to think more concretely about the need for adaptation planning. As a result, we are witnessing increased efforts to develop roadmaps for responding to adaptation governance challenges in both the developed and developing world and, within countries, at multiple levels of governance. As these efforts have evolved, there is an increasing focus on situating adaptation efforts within a resiliency framework, or a framework that focuses on understanding and improving the capacity of systems to cope with disturbances without compromising the core functions and structure of the system. Resilient systems are those capable of encountering disturbances, absorbing change, and recovering in a dynamic fashion. Contextualizing adaptation efforts within a resiliency framework allows policymakers to understand the degree to which different adaptation efforts are needed or viable based on the existing ability of a system to cope with varying levels of disturbance, including climate-related disturbances.

II. Adaptation Planning

Efforts to develop distinct adaptation strategies and more comprehensive adaptation plans are increasing. These efforts include governmentally led efforts at the local, national, and international levels, but they also include a number of initiatives led by private sector and intergovernmental entities. While mitigation efforts to date have approached climate change largely as a collective action problem requiring a centralized, consensus-based top-down decision-making system, there has been early and continuing recognition that adaptation efforts require greater diversification and, often, decentralization of decision-making authority. As leading climate experts have noted, "[a]dapting to climate change involves cascading decisions across a landscape made up of agents from individuals, firms and civil society, to public bodies and governments at local, regional and national scales, and international agencies."[15] Beyond recognizing the growing urgency of adaptation actions and the complex nature of the challenge, both at the domestic and international level, policymakers continue to grapple with the question of how to facilitate something other than a top-down system of adaptation governance or, conversely, fragmentary and uncoordinated local efforts.

In terms of thinking through the character of specific legal responses to adaptation, particularly with regards to natural resource management, there is a vibrant conversation about ways to embed resilience into legal and political responses and exploring the best ways to mandate or incentivize a move towards adaptive management systems. This literature focuses largely on the issue of how "to think about designing legal instruments and institutions now with confidence they will be resilient and adaptive to looming problems as massive, variable, and long term in scale as climate change."[16] The focus is on how to design specific rules and the underlying approaches to structuring specific legal systems, particularly within the context of domestic law. Moreover, as Ruhl describes, three basic strategies have come to dominate adaptation policy planning. These include strategies:

> (1) to *resist* the impacts, also known as the *defend* or *protect* strategy, such as through construction of a sea wall to protect the existing built environment; (2) to increase *resilience* to the impacts, also known as the *adjust*, *accommodate*, or *transform* strategy, such as by changing the form and function of the built environment along the coast to absorb and recover from flooding impacts; and (3) to *retreat* from the impacts by moving away, such as by shifting the built environment away from the coast.[17]

One of the most pressing points of inquiry with respect to adaptation of any kind is the question of at what scale—i.e., local,

---

[15] W. Neil Adger et. al., *Successful Adaptation to Climate Change Across Scales*, 15 GLOBAL ENVTL. CHANGE 77, 79 (2005).

[16] J.B. Ruhl, *Design Principles for Resilience and Adaptive Capacity in Legal Systems—with Applications to Climate Change Adaptation*, 89 N.C. L. REV. 1373, 1373 (2011).

[17] Ruhl, *Climate Adaptation Law, supra* note 4.

regional, national, international—adaptation efforts should occur. In particular, within the U.S. federalist system, questions of scale involve complex questions of jurisdictional rights, transboundary impacts, and sources of funding. While it is evident that the attributes of adaptation make it ill-suited for a top-down, "one size fits all" approach, there is no clear consensus as to when, and with respect to which particular climate risks, adaptation planning and decision making should be devolved to the state or local level and, when devolved, what type of funding and oversight responsibility should be involved.

While there are strong arguments to be made for devolving adaptation efforts to the state and local levels, where understanding of the risk profile of climate threats is greatest, there are also limits to this approach. In particular, state and local authorities may not have access to adequate resources; they may attempt to put their constituencies at an advantage with respect to others with respect to securing scarce resources; the actions they take may impact surrounding communities; and they may not have the ability to coordinate among all affected parties or to maximize potential economies of scale.[18]

Similarly, there are natural benefits and limitations to allocating authority to the federal government. In common with traditional environmental problems, the federal government is well placed to undertake comprehensive scientific investigations and research and development, to identify and address national or transboundary level issues, and to facilitate national consistency in terms of the level of protection that citizens are afforded. On the other hand, the federal government is ill-suited to understand the local and regional threats climate change poses, to identify the most pressing needs, to appreciate local capacity to support and sustain different types of actions, or to create systems that are flexible and are, themselves, capable of adapting as new information becomes available. The tensions associated with the division of authority over adaptation decision making mirror many of the traditional federalist debates over existing systems of environmental law. Ultimately, adaptation requires cooperation among the different levels of government and the more difficult question is how to enable the right actions to occur at the right level. The process of thinking through the division of authority and responsibility for adaptation governance is still very much in its infancy in the United States. Several specific aspects of domestic adaptation governance will be discussed in Part D of this Chapter.

---

[18] *See, e.g.*, Robert L. Glicksman, *Climate Change Adaptation: A Collective Action Perspective on Federalism Considerations*, 40 ENVTL. L. 1159, 1165 (2010).

At the domestic level, traditional questions of jurisdiction are at the center of efforts to develop a more cohesive approach to adaptation. At the global level, the absence of framework for tracking, reviewing, and managing existing adaptation efforts complicates efforts to develop a more coordinated and effective approach to adaptation. That is, despite inevitable tensions between decision makers at the local, state, and federal level, the U.S. federalist system has experience with, and is conducive to facilitating multi-level responses. In the United States, the devil is in the details. In contrast, at the global level, traditional notions of State centrality and the fundamental norm of State sovereignty define the parameters for shaping global approaches to environmental problems. As a result, most multilateral environmental programs function by creating an overarching, centralized institution that establishes a set of goals and obligations that the State parties identify and are then tasked with implementing at the national level. International institutions seek to facilitate compliance with the agreed goals through reporting requirements, financial and technological transfer mechanisms, and, at times, information sharing, but these efforts are generally fragmented and have a mixed track record of success.

Further, these interactions take place between the centralized national government and the Secretariat or Conference of the Parties to the Convention, that is, they are bi-directional from the centralized national government to the centralized international institution. In addition, these interactions are premised on the notions that the centralized international institution is capable of, and has in fact established, an overarching set of rules and that subsequent interactions primarily involve questions about how to comply with those rules in diverse contexts. Accordingly, many existing international environmental institutions struggle to cope with complex environmental challenges, such as adaptation, that require coordinate, multi-level responses and active participation by non-state actors.

III. Adaptation at the International Level

A. The Role of International Governmental Organizations in Adaptation

While there is a need to comprehensively map the field of global adaptation governance, that task is beyond the scope of this chapter. In stepping back to review the adaptation landscape, however, one thing is clear: adaptation efforts are underway in most parts of the world. Adaptation efforts take place in a variety of ways and in a multiplicity of forums.

International adaptation efforts focus on everything from human health, fisheries, forestry, water quality and quantity, coastal zone management, and food security to name but a few key areas. Collective knowledge of efforts across these fields is improving but is still poor. This is due to difficulties in defining what constitutes a climate-based adaptation effort as opposed to a more conventional economic development project, the haphazard way in which these efforts are emerging, the high number of public and private actors facilitating these efforts at multiple levels across the globe, the relatively recent growth of adaptation activity, and the absence of one or more institutions that are tasked with coordinating or tracking climate adaptation efforts.

Recognizing the limits of our knowledge, it is appreciably easier to discern the parameters of national and regional level adaptation strategies as more governmental resources and scholarly attention are directed toward understanding the contours of these systems than to analyze dispersed local actions. Even in key developed and developing nations where national adaptation planning has begun, these "planning efforts are still at a relatively early stage: it is almost more accurate to say that governments are making plans to engage in adaptation planning exercises."[19]

At the local level—particularly in the context of developing countries—one must delve deeper to find the details of planned and ongoing adaptation efforts.[20] In reviewing adaptation planning and guidance documents for specific geographic and issue areas, one common theme is that many of the programs are funded and facilitated by Intergovernmental Organizations (IGOs).[21] Among the most prominent IGOs is the United Nations Development Programme (UNDP). UNDP, together with its partner agencies such as the United Nations Environmental Programme (UNEP), the Food and Agriculture Organization (FAO), the United Nations International Strategy for Disaster Reduction (UNISDR), the World Health Organization (WHO), and the World Bank, facilitate a significant number of ongoing adaptation efforts. Similarly, other regional or thematically focused organizations, such as the World

---

[19] Daniel A. Farber, *The Challenge of Climate Change Adaptation: Learning from National Planning Efforts in Britain, China and the USA*, 23 J. ENVTL. L. 359, 359 (2011).

[20] *See, e.g.*, Rodel D. Lasco et al., *World Agroforestry Centre, Climate Change Adaptation for Smallholder Farmers in Southeast Asia* 5 (2011), http://www.asb.cgiar.org/PDFwebdocs/lasco-2011-ccadaptationfarmerssoutheastasia.pdf; Press Release, UNDP, Launch of Africa Adaptation Programme in Namibia (March 8, 2010), https://collections.unu.edu/eserv/UNU:1403/lasco-2011-ccadaptationfarmers southeastasia.pdf.

[21] *See, e.g.*, Lisa Maria Dellmuth & Maria-Therese Gustafsson, *Global Adaptation Governance: How Intergovernmental Organizations Mainstream Climate Change Adaptation*, 21 CLIMATE POLICY 7 (2021).

Agroforestry Centre, the Asian Disaster Preparedness Center, the Nature Conservancy, and regional and national agencies such as the European Commission, the United Kingdom Department for International Development, and the German Agency for International Cooperation play critical roles in supporting adaptation initiatives worldwide. Each of these groups tends to have a particular geographic area or issue area that they focus on.

The fact that IGOs play a prominent role in adaptation efforts is not surprising given that there is a rich history of IGO involvement in the making and implementation of international environmental law. This remains true in the adaptation context, where demand for specialized assistance is growing rapidly.

With adaptation, the optimal level for adaptation actions varies depending on the particular type of risk the effort seeks to minimize. Given the distinct risk profiles of communities, however, the optimal level for adaptation planning will often be at the local or regional level. Even in those cases where adaptation planning and actions are most ideally taking place at the local level, however, IGOs remain relevant. IGOs have superior access to the resources—for example, technology, finances, and political capital—that are critical to adaptation and often lacking in the areas facing the most acute risk. Without the research, funding, technological and administrative support that they offer, local actors in the most vulnerable parts of the world often lack the resources necessary to adapt to climate change. However, given that creating sustainable adaptation strategies requires the engagement and support of local actors, IGOs can only facilitate not dictate adaptation choices.

While adaptation planning now takes place at virtually every institutional level, these efforts are in the early stages of development, in flux, and unproven. As adaptation becomes a more dominant part of the climate change conversation and as the need for adaptation efforts grows worldwide, one critical question is how the UNFCCC—as the key international climate change institution—alongside other IGOs can help create a framework that informs and facilitates present and future adaptation actions.

B. The Role of the UNFCCC

In the global context, the UNFCCC is the only multilateral institution whose parties are specifically tasked with implementing "measures to facilitate adequate adaptation to climate change" and "cooperat[ing] in preparing for adaptation to the impacts of climate change."[22] As a result, the UNFCCC is an important focal point for

---

[22] May 9, 1992, 1771 U.N.T.S. 107, *reprinted in* 31 I.L.M. 849, art. 4(1) & (2) (1992) [hereinafter UNFCCC].

thinking about the development of adaptation laws and policies. The Paris Agreement advances global efforts to prioritize adaptation efforts, stating that Parties to the agreement "recognize that adaptation is a global challenge faced by all with local, subnational, national, regional and international dimensions, and that it is a key component of and makes a contribution to the long-term global response to climate change to protect people, livelihoods and ecosystems." Thus, the Parties "hereby establish the global goal on adaptation of enhancing adaptive capacity, strengthening resilience and reducing vulnerability to climate change, with a view to contributing to sustainable development and ensuring an adequate adaptation response in the context of the temperature goal."[23] The Paris Agreement further notes the importance of international cooperation and support for adaptation efforts, particularly for developing countries. The only adaptation commitment that the Paris Agreement imposes on parties, however, is to engage in adaptation planning. Thus, in common with the existing system of international climate change law, the Paris Agreement continues to "encourage greater adaptation efforts through softer means."[24]

Even with the Paris Agreement's renewed emphasis on adaptation, neither the UNFCCC nor the Paris Agreement creates a discrete framework—or a set of clear, normative principles—to guide global adaptation efforts. Also, the UNFCCC is not the only international institution whose mission empowers it to facilitate adaptation. Currently, the UNFCCC is not even the lead institution in facilitating adaptation efforts, with institutions such as the UNDP, the FAO, and WHO playing key roles in ongoing adaptation efforts. Nevertheless, the UNFCCC and the Paris Agreement and sit at the center of efforts to facilitate more and more effective adaptation efforts.

As mentioned, while adaptation has always been part of UNFCCC negotiations, until recently it remained a distant second priority to mitigation. In 2010, however, the Parties to the treaty agreed that adaptation must be addressed with the same level of priority as mitigation. At the same time the Parties began the process of constructing a new adaptation framework to facilitate more focused adaptation efforts, including creating a new Adaptation Committee as well as developing a new climate finance mechanism, the Green Climate Fund (GCF). The GCF, which was first proposed as part of the Copenhagen Accord in 2009, was created to "meet the

---

[23] Paris Agreement, Art. 7(1) (Dec. 13, 2015), in UNFCCC, *Report of the Conference of the Parties on its Twenty-First Session*, Addendum, at 21, UN Doc. FCCC/CP/2015/10/Add.1 (Jan. 29, 2016) [hereinafter Paris Agreement].

[24] Daniel Bodansky, *The Paris Climate Change Agreement: A New Hope?*, 110 AM. J. INT'L L. 288, 308 (2016).

financing needs and options for the mobilization of resources to address the needs of developing country Parties with regard to climate change adaptation and mitigation."[25]

Reflecting the shifting priorities of the Parties to the treaty, the governing instrument for the GCF emphasizes the need to balance mitigation and adaptation activities. Following on from commitments first made in 2009 in Copenhagen and subsequently embedded in the Paris Agreement, the Parties to the treaty ultimately agreed that "developed country Parties commit, in the context of meaningful mitigation actions and transparency on implementation, to a goal of mobilizing jointly USD 100 billion per year by 2020 to address the needs of developing countries."[26] As a result, through the GCF, the Parties to the treaty have agreed to mobilize a significant amount of money towards efforts to address climate change, ostensibly with equal parts of that money allocated to adaptation and mitigation.

Efforts to prioritize adaptation within the UNFCCC have highlighted a few emerging characteristics of adaptation policy, namely the importance of these efforts taking place in a multi-level context where the actions are driven from the ground up but are supported by a system of strengthened enabling institutions and the importance of adaptation actions being individualized yet sharing certain common features. More specifically, adaptation actions should be transparent, gender-sensitive and participatory; decisions should be made based on the best-available science; measures should be responsive both to traditional and indigenous knowledge, and to the needs and capacities of vulnerable groups and systems.

The progress made over the past half decade to develop a strategy for climate change adaptation efforts demonstrates that adaptation has risen to the top of the global political agenda. It also reveals that there is an underlying if unstated recognition that adaptation governance is better suited to global facilitation rather than global centralization. That is, ongoing efforts to create an institutional adaptation strategy emphasize the role that the UNFCCC can play in facilitating the creation and implementation of adaptation efforts that are country-owned and county-driven and that are scalable and flexible.

---

[25] United Nations Framework Convention on Climate Change Conference of the Parties, Cancun, Mexico, Nov. 29-Dec. 10, 2010, *Report of the Conference of the Parties on its Sixteenth Session, held in Cancun from 29 November to 10 December 2010*, FCCC/CP/2010/7/Add.1, para. 101 (March 15, 2011).

[26] United Nations Framework Convention on Climate Change Conference of the Parties, CFCCC/CP/2013/L.13, *Draft decision -/CP.19, Work Programme on Long Term Climate Finance*, ¶ 3 (2013), http://unfccc.int/resource/docs/2013/cop19/eng/l13.pdf.

The evolving UNFCCC institutions and, in particular, the Paris Agreement, create a framework for supporting these decentralized strategies not only through traditional—but improved—provisions of financial assistance and technology transfer but also by pushing for improved communication and coordination amongst key U.N. institutions as well as between U.N. and non-U.N. organizations. As it presently exists, the UNFCCC adaptation system, therefore, bypasses a traditional, centralized rules-based system in favor of an institutional framework focused on incentivizing and enabling adaptation efforts at the most appropriate level of governance, which often means at a local or regional level.

The emerging contours of the system suggest that the goal is to enable specific actions to be taken at the optimal level of governance with optimal levels of external support. It is, however, still too early to analyze how the emerging UNFCCC system will be operationalized and whether it will be designed to facilitate interactions between relevant governmental, civil society, and institutional actors.

The decisions of the Conference of the Parties between 2009–present emphasize that the institutional role of the UNFCCC in the context of adaptation is facilitative. One of the most important questions moving forward concerns how the UNFCCC can maximize its ability to facilitate adaptation efforts. The UNFCCC often will not be best situated to offer technical expertise on the details of adaptation strategies or projects in particular fields, e.g., food security or sea level rise. These questions often will be best tasked to the more specialized IGOs, including, for example, the FAO, WHO, UNDP, UNEP, and UNISDR. Instead, the UNFCCC is more aptly positioned to: (1) facilitate coherent national adaptation planning; and (2) coordinate the implementation of adaptation strategies by linking facilitating institutions with relevant actors at the domestic level and providing resources through the evolving Green Climate Fund.

Another important step in harmonizing international efforts was taken with the establishment of the Glasgow Sharm el-Sheikh work program on adaptation established at COP 26 in Glasgow. The program was established to define the future of adaptation efforts ahead of COP 27. The agreement seeks to increase funding for developing countries and build on the progress made at COP 26, where eleven countries and the Belgian region of Wallonia pledged a combined $413 million to the Least Developed Countries Fund. The Glasgow program seeks to increase global awareness and understanding of adaptation, guiding participating countries to implement adaptation programs and communicate the efficacy of these strategies to other participating nations.[27]

The global community is progressing in its efforts to create an institutional framework for national adaptation planning in developed countries, particularly in the least developed countries. The parties to the UNFCCC, through the framework of the Paris Agreement, are also progressing with efforts to create a more effective climate financing mechanism through the Green Climate Fund. Moving forward, climate financing will play a crucial role in the shape and success of adaptation efforts.

C. International Climate Finance

Improving systems of climate finance is fundamental to ongoing efforts to adapt to climate change. Despite more than three decades of efforts, both substantive commitments to climate mitigation and adaptation, and the financing of these efforts remain woefully inadequate. Moreover, the ongoing global pandemic is putting pressure both on political commitments to, and the financing of climate efforts due to strains on governmental resources and widespread economic downturn. As Paul Rose suggests "the economic impact from COVID-19 will reverberate in the economy for years to come,"[28] with far reaching impacts across governmental programs and services.

The question of how to finance efforts to mitigate and adapt to climate change is politically fraught, but there is little doubt that this is an area that has seen rapid growth.[29] The Climate Policy Initiative estimates that, in 2019, "annual [financial] flows rose to USD 579 billion, on average, over the two-year period of 2017/2018, representing a USD 116 billion (25%) increase from 2015/2016. The rise reflects steady increases in financing across nearly all types of investors."[30] Similarly, the Climate Bonds Initiative, which tracks bond issuances labelled as "green" or "climate" bonds, calculates a total of $257.5 billion issued in 2019, with $192.9 issued in 2020, thus

---

27 United Nations Framework Convention on Climate Change, *Glasgow-Sharm el-Sheikh Work Programme on the Global Goal on Adaptation*, https://unfccc.int/sites/default/files/resource/cma3_auv_4ac_Global_Goal.pdf.

28 Paul Rose, *Toward a National Resilience Fund*, NW. U. L. REV. COLLOQUY (2021).

29 *See, e.g.*, Cinnamon Carlarne & David Driesen, *Climate Change Finance*, in THE RESEARCH HANDBOOK ON CLIMATE CHANGE MITIGATION LAW (Leonie Reins & Jonathan Verschuuren eds., forthcoming with Edward Elgar, 2022).

30 *Global Landscape of Climate Finance 2019*, Climate Policy Initiative Report (October 2019), https://www.climatepolicyinitiative.org/publication/global-landscape-of-climate-finance-2019/.

far, with this estimated to increase to $350 billion by the end of the year.[31]

Despite this steady increase, the gap between climate finance needs and climate finance flows remains significant. For example, a 2016 report produced on behalf of the United Nations Environment Programme (UNEP) cautioned that:

> [T]he costs of adaptation are likely to be two-to-three times higher than current global estimates by 2030, and potentially four-to-five times higher by 2050. Previous global estimates of the costs of adaptation in developing countries have been placed at between US$70 billion and US$100 billion a year for the period 2010–2050. However, the national and sector literature surveyed in this report indicates that the costs of adaptation could range from US$140 billion to US$300 billion by 2030, and between US$280 billion and US$500 billion by 2050.[32]

In addition to the adaptation gap, another challenge in climate finance involves determining how much of the monies that are made available are 'new and additional' sources of financing. That is, it is extremely difficult to determine how much of what States are calling 'climate finance' is new and additional to other sources of overseas development aid that was or would have been provided to developing countries under other labels or other programs even absent calls for climate financing and, thus, cannot be treated as flows of finance that are designed to address climate change, separate and additional to other concerns.

Moreover, recent political shifts combined with the ongoing pandemic threaten to derail the improved, but still anemic climate finance commitments from governmental actors. For example, as Sachs points out:

> Developed nation pledges to the Green Climate Fund, the leading source of governmental climate assistance to the developing world, have totaled $10.3 billion since 2010, but governments have transferred only about $3.5 billion to the Fund. In 2017, President Trump terminated any new U.S. contributions to the Fund. On the verge of the new decade, it seems unlikely that the developed world will raise and distribute $100 billion *annually* through the

---

[31] *The Climate Bond Initiative*, https://www.climatebonds.net.

[32] Daniel Puig et al., *The Adaptation Finance Gap Report 2016. United Nations Environment Programme* (2016), https://unepdtu.org/publications/the-adaptation-finance-gap-report/.

2020s to finance climate change mitigation and adaptation in the developing world.[33]

Even if flows of funding increased to desired levels of $100 billion per year and, even if this funding was 'new and additional', existing estimates suggest that much more is needed. A recent UNEP report suggests that "annual adaptation costs in developing countries alone are estimated at USD 70 billion. This figure is expected to reach USD 140–300 billion in 2030 and USD 280–500 billion in 2050. An increase in financing will be critical for countries to meet their adaptation goals."[34]

While estimates of need vary, several aspects of adaptation financing are clear: costs are increasing; public funding is increasing; there remain substantial gaps between needs and available funds; the majority of climate financing still goes to mitigation; those funds that are available for adaptation are hard to track both in terms of how they overlap with other funding programs and in terms of how they are spent.

In addition to inadequate levels of finance, there are other important impediments to improving climate finance systems. One of the primary challenges hindering more effective climate financing is the structural fragmentation of finance mechanisms.

Climate finance is highly fragmented. Under the UNFCCC, alone, a handful of mechanisms exist or are used to finance mitigation and adaptation measures, including: the Global Environmental Facility (GEF), the Adaptation Fund, the Least Developed Countries Fund, the Special Climate Change Fund, and the Green Climate Fund.

The GEF was the original funding entity for the UNFCCC, however, between 2000–2001, the Adaptation Fund, the Special Climate Change Fund, and the Least Developed Country Fund were created as additional funding mechanisms. All three of these funds are designed to provide additional assistance to developing countries for adaptation or technology transfer and economic diversification. Except for the Adaptation Fund[35], however, these funds are based on voluntary contributions and the resources that they have, and that they have been able to disperse, are extremely limited when compared to estimations of need.

---

[33] Noah M. Sachs, *The Paris Agreement in the 2020s: Breakdown or Breakup*, 46 ECOLOGY L.Q. 865, 898 (2019).

[34] UNEP, *The Adaptation Gap Report 2020*. United Nations Environment Programme: Executive Summary xii (2021), https://www.unep.org/resources/adaptation-gap-report-2020.

[35] The Adaptation Fund is partially financed by a 2 percent levy on all Certified Emission Reduction Units produced by Clean Development Mechanism projects.

Each of these mechanisms, as well as the previously discussed Green Climate Fund, possess different missions, leadership, and operating policies meaning that they intersect and, to some extent, compete with one another to receive and funnel North-South finance, creating a fragmented and ill-coordinated climate finance regime. Further complicating the climate finance scene, in addition to the UNFCCC-linked financing mechanisms, there are a host of other financing institutions, ranging from international bodies such as the International Monetary Fund (IMF), to regional development banks, to individual state aid programs, to non-profits that are working in this area and serving as conduits for climate financing. As a result, the larger field of climate finance, as well as the narrower field of adaptation finance, is simultaneously complex and piecemeal. The variety of funding sources allows experimentation and variety in terms of the types of adaptation efforts that are supported, but it also makes it difficult to evaluate the efficacy of different strategies and the extent to which adaptation needs are being addressed.

As adaptation financing becomes an increasing focal point of the climate change debate, financing entities are forced to grabble with difficult questions about how to decide how to allocate limited pools of funding. Decisions about how to structure financing mechanisms implicate important questions of equity that lie at the heart of climate change. Climate change creates equity dilemmas because of the way in which it operates. In particular, there are disconnects both in place and in time between the causes and effects of climate change. That is, the distribution of climate-related risks and impacts is "independent of [the] emissions profile of each nation, and . . . the impacts are felt over a long time horizon due to the long life of greenhouse gases in the atmosphere."[36] In other words, the climate risks that each country faces are not proportional to the amount of greenhouse gas emissions that the country produces. From an adaptation perspective, one of the most troubling challenges is that many of the countries at greatest risk from climate change are also those countries that both have the least capacity to adapt and contributed least to the problem of anthropogenic climate change.

Equity concerns are not unique to the adaptation conversation. The UNFCCC acknowledges and seeks to rectify equity concerns across the spectrum of climate-related policies relying on the concept of "common but differentiated responsibilities and respective capacities" (CBDR), as discussed in Chapter 3. As the ethical cornerstone of the UNFCCC, the principle of CBDR has proved controversial because how this concept is understood and applied

[36] P.R. Shukla, *Justice, Equity and Efficiency in Climate Change: A Developing Country Perspective, in* FAIRNESS CONCERNS IN CLIMATE CHANGE (Ference Toth Ed., 1999).

affects how the collective burden of climate change obligations is distributed. In the context of adaptation planning and financing, evolving understanding of CBDR shapes how we think about who should be contributing to adaptation efforts and how available resources should be allocated.

These questions become increasingly important as the Green Climate Fund, and other financing bodies, begins to mobilize and distribute increasing amounts of public and private finance. At the moment, with regard to private finance, little consideration is paid to the role that equity should play in adaptation financing. As Miles explains:

> Currently, there is no requirement for transnational financing and investment decision-making to be supportive of sustainable development. There is no international accountability for financing socially and environmentally harmful projects. And there is no international regulatory framework that can require equitable private sector financing of climate adaptation measures in developing states.[37]

With regards to monies funneled through the UNFCCC and other public institutions, each funding mechanism has different decision-making processes for allocating funds, but early evidence suggests that equity considerations play a limited role in influencing lending decisions. For example, of the top twenty countries approved for funding through the Adaptation Fund—a Fund specifically created to fill a funding gap for developing country Parties to the Kyoto Protocol that are particularly vulnerable to the adverse effects of climate change—only four are in the top fifty "at-risk" countries ranked by the Germanwatch's Global Climate Risk Index (1994–2013). The fact that the Adaptation Fund focuses on developing countries suggests that equity is an intrinsic part of the lending process, since all potential recipient countries are developing countries and, thus, are countries that bear little responsibility for contributing to climate change while also standing to carry a disproportionate burden when it comes to suffering the effects of climate change. But this framing of equity fails to account for the great variation of both need and capacity among developing States that contributes to climate change risk profiles and resulting adaptation inequalities.

The Green Climate Fund is still in an emergent phase, so it is not yet possible to determine the ways in which it will reshape adaptation efforts. The Governing Instrument for the Fund

---

[37] Kate Miles, *Investing in Adaptation: Mobilising Private Finance for Adaptation in Developing States*, 5 CARBON & CLIMATE L. REV. 190, 201 (2011).

emphasizes the need to balance mitigation and adaptation funding and to promote promoting environmental, social, economic and development co-benefits of financing decisions while taking into account the needs of countries that are particularly vulnerable to climate change. The steps that are being taken to ensure that adaptation funding is both equally prioritized with mitigation funding, and more evenly distributed suggests that the Fund is being developed in such a way as to elevate adaptation funding and minimize the equity imbalances that have plagued other UNFCCC programs.

Questions of what equity means, and what role different understandings of equity play in adaptation financing decisions take on additional importance as climate change adaptation efforts begins to intersect with other areas of law, namely disaster law.

D. Climate Change and International Disaster Response

Efforts to develop effective approaches to adaptation policy take on additional urgency as our understanding of the links between climate change and disasters deepens. Worldwide, existing systems for disaster preparedness, response, and mitigation are already stressed. In many cases, climate change heightens disaster risk and adds further stress to these systems. Adapting to climate change, therefore, requires understanding and improving disaster preparedness, mitigation, and response systems at every level. While our understanding of disaster-climate linkages is improving, there is still a need to deepen understanding of the ways in which climate change exacerbates the risks associated with both sudden disasters (e.g., storms and flooding) and slow-onset disasters (e.g., drought and famine).

On the financing side, there are opportunities to maximize existing pools of adaptation and disaster funding by focusing on projects that seeks to achieve both adaptation and disaster risk reduction goals. In common with adaptation financing, data on how much disaster risk reduction financing there is and how these funds are spent is deficient. The data that is available, however, reveals the scale of disasters and the importance of understanding the linkages between climate change adaptation and disaster response. The UN Office for Disaster Risk Reduction, for example, estimates that "over the last twenty years, 7,348 disaster events were recorded worldwide" and that these disasters "claimed approximately 1.23 million lives, an average of 60,000 per annum, and affected a total of over 4 billion people" and "led to approximately US$ 2.97 trillion in economic losses worldwide."[38] Disasters are a major cause of loss of life, human suffering, and economic loss and current levels of financing cannot keep pace from a prevention, response, or recovery perspective. Because climate change often exacerbates disaster-related threats, understanding and responding to the linkages between climate change and disasters is essential to minimizing social and economic costs and to maximizing the efficiency of efforts to address both climate change and disasters.

In common with adaptation, efforts to prioritize disaster risk reduction efforts are gaining momentum at every level of governance. In recent years, for example, the international community has prioritized disaster risk reduction efforts and disaster and climate hot spot countries, such as Indonesia and the Philippines, have begun to invest heavily in disaster risk reduction efforts. Moving forward, disaster risk reduction and response efforts are likely to be increasingly viewed as complementary to climate change adaptation efforts and vice versa.

IV. Adaptation at the Domestic Level

The international system is largely aimed at financing and supporting adaptation efforts at the national level. In this section, we turn the focus to U.S. adaptation efforts.

A. An Overview of Adaptation Measures

In the United States, in contrast to mitigation, where there is an emerging legal infrastructure for reducing greenhouse gas emissions from primary sources, there is very little in terms of formal, legal rules guiding adaptation efforts at the federal level. As Gerrard describes it, with respect to adaptation law in the United States, "it is striking how little hard law there is that is explicitly aimed at adaptation, at increasing resilience, or at reducing vulnerability to climate change."[39] Instead, in the United States, the federal government has largely played a facilitative role to date. This facilitation has included developing a better example of what types of threats climate change poses nationwide and what types of responses are possible, as well as providing financial resources for states and localities engage in climate change adaptation efforts.

Pursuant to the Global Change Research Act of 1990, the U.S. Global Change Research Program (USGCRP), a multi-agency group consisting of representatives from thirteen federal departments and agencies, is tasked with preparing and submitting to the President

[38] UN Office for Disaster Risk Reduction, *Human Costs of Disasters: An Overview of the Last 20 Years, 2000–2019* (2020), https://www.undrr.org/publication/human-cost-disasters-overview-last-20-years-2000-2019.

[39] *See* Gerrard, *supra* note 1, at 11.

and Congress assessments of the effects of global change in the United States. The report is intended to facilitate legal and policy-making decisions at the federal and sub-federal levels and, thus, provides a comprehensive view of climate change sources and threats, including anticipated climate impacts by U.S. regions and sectors.

The most recent Assessment was released in 2018.[40] The 2018 report, spanning more than 1500 pages, summarizes the wide variety of threats and potential responses to climate change, noting, in key part that "[c]limate change creates new risks and exacerbates existing vulnerabilities in communities across the United States, presenting growing challenges to human health and safety, quality of life, and the rate of economic growth," and cautioning that "without substantial and sustained global mitigation and regional adaptation efforts, climate change is expected to cause growing losses to American infrastructure and property and impede the rate of economic growth over this century."[41] Having noted the centrality of adaptation efforts, the report notes that "adaptation planning and implementation activities are occurring across the United States in the public, private, and nonprofit sectors" but are "not yet commonplace."[42]

The report then highlights specific adaptation options that are being pursued in the United States. At the local level, for example, following Superstorm Sandy, New York City has taken steps to reduce the impact of future floods, including by relocating households out of the most flood-prone areas, raising the height of certain structures above the ground so they suffer less damage from future flooding, and training the officials responsible for revising building codes and land-use policies to use the most up-to-date estimates of flood risk.[43] Similarly, Miami Beach is raising the levels of its roads and constructing seawalls in anticipation of increasingly high sea levels.

Another effort at the local level is Philadelphia's efforts to minimize heat-related mortality. Facing warming conditions, in the mid-1990s, Philadelphia adopted a system for reducing the risk of death during heat waves. Philadelphia's efforts include creating a "Heatline", staffed by nurses, developing a "Cool Homes Program" that helps retrofit the homes of those most vulnerable to heat, and generally improving heat-related warning systems and response infrastructure. Demonstrating the value of local level responses, it is

---

40 *See Fourth National Climate Assessment, supra* note 14.

41 *Id.*

42 *Id.* at 1310.

43 *Id.* at 1320.

estimated that Philadelphia's system has led to a decline in heat-related mortality rates.[44]

With respect to specific sectors, the transportation sector is a focus area for adaptation efforts across the United States. In the Gulf Coast, for example, there is a single road, Louisiana Highway 1, that connects Port Fourchon, Louisiana with the rest of the United States. Port Fourchon, in turn, supports 75% of deepwater oil and gas production in the Gulf of Mexico. The road into Port Fourchon is, therefore, a vital route that allows for the movement of workers, supplies, machinery between the mainland and Gulf oil operations. Similarly, in times of emergency, Louisiana Highway 1 is the only evacuation route available to both onshore and offshore workers. The Gulf region, including Highway 1, is increasingly vulnerable to storm surge and flooding. As a result, the state of Louisiana is investing in efforts to upgrade Highway 1 so that it is more resilient to changing conditions.

Similarly, in the US- affiliated Federated States of Micronesia, are planning a "climate-proof" road around one of the islands that would provide access to a remote village, despite the added costs associated with "climate-proofing" the road to withstand projected increases in heavy precipitation and sea-level rise. Micronesian authorities concluded that the investment would pay off in the long-run due to ongoing and projected climate changes.

B. The Federal Role

These examples are just a few of the ongoing adaptation efforts that the Assessment highlights. What the report reveals is that adaptation efforts are taking place nationwide and multiple levels of governance and across key economic sectors, but these efforts tend to be piecemeal, fragmented, and small in scale compared to the anticipated impacts of projected climatic changes.

In 2009, Obama brought more attention to these challenges when he convened a new task force to explore climate adaptation needs and issued Executive Order 13514, Federal Leadership in Environmental and Energy Performance, that directed federal agencies not only to takes steps to mitigate their impact on climate change, but also to develop agency-specific climate adaptation plans. Federal Agencies began releasing their first Climate Adaptation plans in 2013; the agency-specific plans outlined "strategies to reduce the vulnerability of Federal programs, assets, and investments to the impacts of climate change, such as sea level rise or more frequent or

---

[44] Kate R. Weinberger et. al, *Effectiveness of National Weather Service Heat Alerts in Preventing Mortality in 20 US Cities*, 116 ENVIRONMENT INTERNATIONAL 30 (2018); City of Philadelphia, *Safety and Emergency Preparedness*, https://www.phila.gov/services/safety-emergency-preparedness/natural-hazards/excessive-heat/.

severe extreme weather. Agency plans highlight actions to plan for and address these impacts in their programs and operations."[45]

The task force that President Obama created, the Interagency Climate Change Adaptation Task Force, was co-chaired by the Council on Environmental Quality (CEQ), the Office of Science and Technology Policy (OSTP), and the National Oceanic and Atmospheric Administration (NOAA) and included representatives from more than 20 Federal agencies. It was tasked with providing recommendations on how "[f]ederal policies, programs, and planning efforts can better prepare the United States for climate change."[46]

In October 2010, the Task Force recommended a set of policy goals and actions in its Progress Report to the President focusing on how the Federal Government could coordinate with and work with partners at the local, state, and tribal levels. In key part, the Task Force's recommendations focused on five key areas: integrating adaptation into federal government planning and activities; building resilience to climate change in communities; improving accessibility and coordination of science for decision making; developing strategies to safeguard natural resources in a changing climate; and enhancing efforts to lead and support international adaptation. The emphasis on coordination revealed the degree to which adaptation requires a varied, multi-layered approach, with significant emphasis placed on local and regional level actions but with those actions supported by scientific, regulatory, and, at times, economic resources from the federal level. The emphasis on international adaptation reflects growing concern about how climate change can exacerbate existing systemic vulnerabilities leading not only to local and regional harms, but also international security threats.

Through the Task Force, the Executive Order, and complementary efforts, President Obama sought to mobilize Federal resources to facilitate the integration of adaptation concerns into decision-making processes at every level of governance. These efforts prompted positive change by integrating adaptation into federal agency planning processes, by encouraging more research into adaptation needs and strategies, and by facilitating more expansive coordination between federal, state, and local adaptation efforts.

During the Trump Administration, federal adaptation efforts were curtailed. The Trump Administration sought to purge climate

---

[45] The White House, President Barrack Obama, Council on Environmental Quality, *Climate Resilience,* https://obamawhitehouse.archives.gov/administration/eop/ceq/initiatives/resilience.

[46] Progress Report of the Interagency Climate Change Adaptation Task Force, *Federal Actions for a Climate Resilient Nation* at iv, Oct. 28, 2011, https://obamawhitehouse.archives.gov/sites/default/files/microsites/ceq/2011_adaptation_progress_report.pdf.

change considerations from natural resource management strategies, including revoking an Obama-era executive order promoting "climate resilience" in the Bering Sea[47] and rescinding an Obama-era policy that integrates climate change into natural resource management decisions in national parks.[48] The Administration also rescinded an Obama-era policy directing the Department of Interior to "integrate climate change adaptation strategies into its policies, planning, programs and operations"[49] The cumulative effect of the Trump Administration's roll-backs of resiliency and adaptation policies was to minimize the extent to which the federal government must or even can take climate change into account when making short and long-term planning decisions across a range of issues.[50] These roll-backs undermined what was already a thin and experimental set of strategies that President Obama had put in place to try to anticipate and respond to the pervasive threats that climate change poses to the United States. In common with efforts worldwide, U.S. adaptation planning is still in its infancy. Eroding the emerging foundations for national adaptation policy set the United States back with the effect of minimizing the federal government's ability both to limit the negative effects and to take advantage of any short-term positive effects of climate change.[51]

The Biden Administration has begun the process of reversing the Trump Administration's rollbacks. For example, the Administration has prioritized federal agency climate adaptation

---

[47] *See* Exec. Order No. 13,795, 3 C.F.R. 340 (2017).

[48] *See* Rob Hotakainen, *NPS Chief Scraps Climate-Focused Order*, E&E NEWS (Aug. 31, 2017), https://www.eenews.net/stories/1060059511 [https://perma.cc/9UFC-EEWH].

[49] *See* Michael Doyle, *Department Rescinds Obama-era Mitigation and Climate Docs*, E&E NEWS (Jan. 5, 2018), https://www.eenews.net/stories/1060070247 [https://perma.cc/C3RD-ZZB7].

[50] For more details on the Trump era roll backs, *see Climate Change, Sustainable Development, and Ecosystems, in* ABA SECTION ENV'T, ENERGY, & RESOURCES*: 2017 ANNUAL REPORT* 339–40 (Andrew Schatz et al. eds., 2017); STATE ENERGY & ENVTL. IMPACT CTR., NYU SCH. L., STATE ATTORNEYS GENERAL: 13 MONTHS OF CRITICAL ACTIONS 21–22 (Feb. 2018), https://gallery.mailchimp.com/8c3272f6ebbb6024dc1359725/files/fdbd6457-5cff-4672-8bd7-5cae63ba69aa/Web_Report_StateImpactCenter_Final.04.pdf [https://perma.cc/39LG-7LTQ].

[51] *See, e.g.*, J.B. Ruhl, *The Political Economy of Climate Change Winners*, 97 MINN. L. REV. 206, 247, 269–70 (2012) (exploring the reality that, even if the global aggregate impacts of climate change are negative, some people—and some groups of people—stand to benefit from climate change in the near term, and discussing the complex interplay between climate change winners and losers over time); *see also* Robin Kundis Craig, *The Social and Cultural Aspects of Climate Change Winners*, 97 MINN. L. REV. 1416, 1417, 1418, 1420 (2013) (cautioning that how we label people who benefit during times of social turmoil "depends as much on cultural constructions of their meaning and public relations as on actual differences in their motives and actions" and warning that "winners could come at the expense of ultimately disastrous long-term consequences for the planet as a whole").

and resiliency planning efforts by having all large federal agencies develop adaptation and resiliency plans. As described by the Administration, the plans are designed to "leverage procurement decisions to drive innovation and increase resilience against supply chain disruptions and deliver on President Biden's commitment to invest climate and clean energy dollars in environmental justice communities."[52]

Moreover, as described in Chapter 5, the $1.2 trillion Infrastructure Act (2021) is designed to help advance the Biden Administration's prioritization of adaptation and resiliency planning through massive investments in physical infrastructure (e.g., ports, airports, freight) to make it more sustainable and resilient, as well as programs to improve community resiliency through investments in weatherization of homes and other efforts to protect against drought, heat, and floods.[53] As one example, through Infrastructure Act funding, in 2022 the U.S. Department of Agriculture invested more than $100 million in reforestation to help mitigate the impacts of climate change and rebuild forested areas in the aftermath of wildfires.[54] The Biden Administration efforts were still in their early stages at the time of writing but the renewed emphasis on adaptation coupled with massive new flows of funding through the Infrastructure Act create a more solid foundation than previously existed for sustained federal adaptation and resiliency efforts.

C. State and Local Efforts

As the America's Climate Choices report highlights, however, actions at the sub-federal level are increasing in number, scale, and intensity. While, in common with Federal efforts, these efforts remain similarly diffuse and fragmented, they reveal a trend towards increased concern about climate impacts at the local, state, and regional levels and increased willingness to leverage legal and economic resources towards minimizing risks and maximizing resiliency. For example, as of 2022, eighteen states had completed adaptation plans, another five were in the process of developing an

---

[52] The White House, *Fact Sheet: The Biden Administration Releases Agency Climate Adaptation and Resilience Plans from Across Federal Government* (Oct. 7, 2021), https://www.whitehouse.gov/briefing-room/statements-releases/2021/10/07/fact-sheet-biden-administration-releases-agency-climate-adaptation-and-resilience-plans-from-across-federal-government/.

[53] The White House, *Fact Sheet: The Bipartisan Infrastructure Deal Boosts Clean Energy Jobs, Strengthens Resilience, and Advances Environmental Justice*, (Nov. 8, 2021), https://www.whitehouse.gov/briefing-room/statements-releases/2021/11/08/fact-sheet-the-bipartisan-infrastructure-deal-boosts-clean-energy-jobs-strengthens-resilience-and-advances-environmental-justice/.

[54] U.S. Department of Agriculture, *Biden-Harris Administration Announces Plans for Reforestation, Climate Adaptation, including New Resources from Bipartisan Infrastructure Law* (July 25, 2022), https://www.usda.gov/media/press-releases/2022/07/25/biden-harris-administration-announces-plans-reforestation-climate.

adaptation plan, and all but eleven states have local and/or regional plans in place.[55]

Among these state-level actions, coastal states have been at the forefront of leading adaptation planning. Facing the likelihood of severe climate-related risks, including sea level rise, extreme events (storms, droughts, fires), and natural resource stress, states such as California and Hawaii are leading efforts to develop comprehensive state adaptation plans and strategies.

California's efforts, for example, date back to 2005, when then-Governor Schwarzenegger established a Climate Action Team (CAT) to begin coordinating state-level climate responses. Part of CAT's responsibilities was drafting biennial science assessment reports exploring the climate impacts and adaptation options for California. Building on this initiative, in 2008, Governor Schwarzenegger, by Executive Order, directed state agencies to begin planning for sea-level rise and climate impacts. Pursuant to this Order, in 2009, California adopted its first statewide Climate Adaptation Strategy (CAS). The CAS summarizes the impacts of climate change on California and recommends adaptation strategies across seven key sectors: public health, biodiversity and habitat, oceans and coastal resources, water, agriculture, forestry, and transportation and energy.[56] The 2009 CAS was innovative in that it "was the first of its kind in the usage of downscaled climate models to more accurately assess statewide climate impacts as a basis for providing guidance for establishing actions that prepare, prevent, and respond to the effects of climate change."[57]

California has continued to build on these early adaptation efforts through the release of updated climate assessments, the creation of an Adaptation Planning Guide that facilitates local level adaptation planning, and the development of the Cal-Adapt website,[58] which allows users to understand and plan for local climate related risks. Equally important, recognizing the links between climate impacts and hazard or disaster-related risks, California has also created, and linked to adaptation data, natural hazard data and planning tools.[59] Finally, California also supports a

---

[55] *See* The Center for Climate and Energy Solutions, *State and Local Climate Adaptation,* https://www.c2es.org/us-states-regions/policy-maps/adaptation; *See* Georgetown Climate Center, *States Adaptation Progress Tracker,* https://www.georgetownclimate.org/adaptation/plans.html.

[56] CA.gov, California Climate Change, *California Adaptation Strategy,* https://resources.ca.gov/Initiatives/Building-Climate-Resilience/2021-State-Adaptation-Strategy-Update.

[57] *Id.*

[58] Cal-Adapt, http://beta.cal-adapt.org/.

[59] *See* CalOES, *My Hazards,* http://myhazards.caloes.ca.gov/.

number of discrete efforts to facilitate local level adaptation efforts through financial assistance.

In common with California, Hawaii has been ramping up statewide adaptation efforts. Following earlier failed attempts to jumpstart state adaptation planning, in 2012, Hawaii passed a law incorporating prioritizing adaptation planning and integrating climate adaptation strategies into land-use in planning policies.[60] Subsequently, Hawaii established, by law, an Interagency Climate Adaptation Committee, authorized the Hawaii Department of Land and Natural Resources and Office of Planning to create a Sea Level Rise Vulnerability and Adaption Report by December 2017, and tasked the Office of Planning with coordinating the development of statewide climate adaptation plans. In 2017, Hawaii adopted a law endorsing the goals of the Paris Agreement and creating the Hawaii Climate Change Mitigation and Adaptation Commission. The law tasked the new Commission with leading state efforts to limit greenhouse gas emissions and improve resiliency, in line with the objectives of the Paris Agreement.[61] In relevant part, the Commission was tasked with assessing vulnerability to sea level rise as well as other projected climate impacts, and assessing Hawaii's capacity to meet its adaptation goals, which it is also tasked with setting.

California and Hawaii's efforts to create state-level adaptation policies represent progressive examples of state efforts to facilitate climate change adaptation. However, many other states are beginning to take climate risks seriously and to take steps to limit the expected impacts of climate change, often beginning by focusing on the sectors most at risk in their state—such as agriculture in the mid-west, flood plains along the Mississippi, water resources in the West, and sea-level rise along the East Coast.

States are not the only entities ramping up adaptation efforts. In fact, it is estimated that "roughly 20 percent of cities around the globe have developed adaptation strategies", including more than 100 in the United States.[62] In addition to the efforts by New York City and Philadelphia described above, a number of large and mid-size cities and counties have begun to develop different types of adaptation and resiliency strategies. In 2016, for example, Charleston, South Carolina adopted a Sea Level Rise Strategy to help cope with flooding and sea level rise challenges. In another coastal

---

[60] Hawaii, Act 286, Climate Change Adaptation Priority Guidelines, http://www.capitol.hawaii.gov/hrscurrent/Vol04_Ch0201-0257/HRS0226/HRS_0226-0109.htm.

[61] HI SB 559 (June 8, 2017).

[62] Maria Gallucci, *Six of the World's Most Extensive Climate Adaptation Plans*, INSIDE CLIMATE NEWS (June 20, 2013), https://insideclimatenews.org/news/20130620/6-worlds-most-extensive-climate-adaptation-plans.

area, the counties of Broward, Miami-Dade, Palm Beach, and Monroe entered into the Southeast Florida Regional Climate Change Compact to coordinate regional mitigation and adaptation efforts. More specifically, the Compact "calls for the Counties to work cooperatively in order to develop a coordinated response to proposed state climate legislation and policies; to create a Southeast Florida Regional Climate Change Action Plan to include mitigation and adaptation strategies; and to meet annually in Regional Climate Summits to mark progress and identify emerging issues."[63] In addition, the plan made over 100 "actionable" recommendations that were designed to be completed within the next five years.

Local, state, and regional adaptation efforts, as well as international cooperation on adaptation, are growing and this trend is likely to continue as evidence mounts of the risks that climate change poses to local and regional systems. As one recent analysis confirms, the combined impact of global and local climate change on urban economies is significant and growing with the total economic costs of climate change for cities this century increasing rapidly by the end of the century, and worsening as climate change amplifies urban heat island effects.[64] Moving forward, cities, counties, and states are likely to be the focal point of adaptation and resiliency efforts and innovation as they experience the localized negative effects of climate change and experiment with ways to minimize climate risks and impacts.

D. Adaptation and Property Rights

The U.S. Constitution requires that owners be compensated when their property has been taken by the government. For at least a century, the courts have held that property can be "taken" through regulation as well as outright seizure. One important adaptation strategy is to prevent development in areas that are currently at risk or expected to be so in the future. In addition, current proper uses might be discontinued in order to decrease the level of risk. These adaptation measures could pose liability risks under the taking clause.

The risk of takings liability is probably minimal for zoning changes due to sea level rise, permitting requirements for beach nourishment, or substitution of beach replenishment (rather than

---

[63] Adaptation Clearinghouse, *Southeast Florida Regional Climate Action Plan*, http://www.adaptationclearinghouse.org/resources/southeast-florida-regional-climate-action-plan.html.

[64] Francisco Estrada, W.J. Wouter Botzen, & Richard S.J. Tol, *A Global Economic Assessment of City Policies to Reduce Climate Change Impacts*, 7 NATURE CLIMATE CHANGE 403 (2017).

sea walls) as a defense against rising seas.[65] But other measures are more likely to prompt serious takings claims, including bans on voluntary efforts by property owners to protect their property or prohibitions on permanent structures in areas at risk.[66]

A quick review of takings doctrine may be helpful to understanding these claims. The Supreme Court currently employs three separate tests to determine whether a regulation should be considered a "taking" of property that requires compensation.[67] First, the Court finds a taking when the government mandates a physical intrusion on private property. Such an intrusion is a taking even if it does not cause any harm to the owner, either in economic terms or as an invasion of privacy.[68] The second category, established in *Lucas*, applies to so-called "total takings," where the government has eliminated any possible economically beneficial use of the property.[69] The third category covers the remaining cases. This default category is governed by the *Penn Central* test,[70] which requires a determination of whether the government regulation interferes with reasonable, investment-backed expectations.

Takings doctrine can be a serious problem for a variety of preservation laws, such as those designed to prevent the destruction of wetlands or biodiversity.[71] Such regulations may prevent the development of all or part of an owner's land or may require use of the land for the public to access waterways or other public areas. As those cases illustrate, courts have been resistant to regulations banning all development of property.

Efforts to restrict development in coastal areas due to sea level rise may also face significant takings challenges. But there are some

---

[65] *See* Michael Allan Wolf, *Strategies for Making Sea-Level Rise Adaptation Tools "Takings Proof,"* 28 J. OF LAND USE 157, 175 (2013).

[66] *Id.* at 183.

[67] Takings jurisprudence extends back through the Founding Era, and roots in legal theory are even older, but the current approach to regulatory takings is a relatively modern development. *See* Joseph L. Sax, *Takings and the Police Power*, 74 YALE L.J. 36, 38–60 (1964). One constant since Professor Sax wrote fifty years ago is that "the predominant characteristic of this area of law is a welter of confusing and apparently incompatible results." *Id.* at 37.

[68] Admittedly, this rule has received some justified criticism. *See, e.g., Andrea Peterson, The False Dichotomy Between Physical and Regulatory Takings Analysis: A Critique of Tahoe-Sierra's Distinction Between Physical and Regulatory Takings*, 34 ECOLOGY L.Q. 381 (2007).

[69] The Court recognized an important exception, allowing an activity to be completely banned when it constitutes a common-law nuisance. For discussion of this exception, *see* Richard Lazarus, *Putting the Correct 'Spin' on Lucas*, 45 STAN. L. REV. 1411 (1993).

[70] The test derives from Penn Central Transportation Co. v. New York, 438 U.S. 104 (1978).

[71] *See* John Echeverria and Julie Lurman, *Perfectly Astounding Public Rights: Wildlife Protection and the Taking Clause,* 16 TULANE ENV. L. J. 333 (2003).

ambiguities in current doctrine that might help restrictions on coastal property development avoid the *Lucas* doctrine.[72] For instance, though bans on hard armoring like sea walls may result in some sacrifice to coastal owners, they serve important public purposes.[73] For that reason, we cannot be sure how they would be resolved under current law.

An additional challenge in the context of adaptation and property rights is the phenomenon of shrinking shorelines. There are different doctrines to address gradual and sudden shifts in shorelines, but these doctrines do not address the relationship between rising sea levels, shoreline changes, and literal property boundaries.[74] These ongoing changes will increasingly raise questions over public-private property boundaries.

The takings and shifting-boundary problems are indications of the challenge that climate change will pose to institutions such as property that take for granted an unchanging physical backdrop for human activities. One of the key functions of property law is to uphold the reasonable expectations of owners. As we move into an era where people become more and more accustomed to climate change, their "reasonable expectations" may shift to acknowledge the risks that some land uses may no longer be tenable or safe.

V. The Future of Adaptation Law

Adaptation law at both the international and domestic levels remains thin and fragmented. At the international level, the Paris Agreement advances efforts to place adaptation efforts on par with mitigation efforts, to facilitate improved adaptation planning, and to increase flows of money towards adaptation efforts. These efforts, however, are still young and existing adaptation plans are fragmented, poorly implemented, and under financed. Increasingly, however, international institutions are emphasizing the importance of limiting climate risks and are recognizing the linkages between disaster and adaptation planning. As efforts continue to implement the objectives of the Paris Agreement and to operationalize the Green Climate Fund, international efforts to support adaptation are likely to grow, but to remain partial and fragmented.

At the domestic level, the federal legal adaptation architecture is extremely thin, but is being bolstered through executive actions and the 2021 Infrastructure Act. In contrast, adaptation efforts are proliferating rapidly at the city, county, and state levels. As cities

---

[72] *See, e.g.*, Wolf, *supra* note 65, at 184–195.

[73] *See* Comment, *Fix It or Forget It: How the Doctrine of Avulsion Threatens the Efficacy of Rolling Easements*, 51 HOUSTON L. REV. 297, 302–307 (2013).

[74] *See Ruhl, Climate Adaptation Law, supra* note 4, at 12.

and states experience the negative impacts of climate change, this pattern is likely to continue.

Although we have focused on governmental actors, private actors will be heavily involved in climate adaptation. Many firms have facilities in coastal areas or inland flood plains. Firms may be under increasing pressure from investors, creditors, and insurers to engage in adaptation. They may also demand greater disclosure of potential risks. Indeed, failure to engage in effective adaptation might be a basis for liability if harm to third-parties results. Insurance companies will have to reevaluate their risk projections and premium levels in light of changing risks. Temperature changes and precipitation changes will require farmers to change crops or adopt new methods of irrigation or flood control. Individual homeowners will have to consider escalated risks from flooding in some places and wildfires in others.

Moving forward, the level of adaptation response needed globally and locally will depend on the level and success of mitigation efforts. While, some level of negative climate impacts is inevitable, the extent of these impacts is determined, in part, by ongoing mitigation responses. As a result, while mitigation and adaptation efforts require different legal and political responses, they are intertwined and successful efforts to address climate change require strong mitigation and adaptation responses.

Further Readings

Cinnamon Carlarne, *Rethinking a Failing Framework: Adaptation and the Future of the Global Climate Change Regime,* 25 GEORGETOWN INTL ENVTL L. R. 1 (2013).

Robin Kunis Craig, "*Stationarity is Dead*"—*Long Live Transformation: Five Principles for Climate Change Adaptation Law*, 34 HARV. ENVTL. L. REV. 9 (2010).

Daniel A. Farber, *Adapting to Climate Change: Who Should Pay?*, 23 J. LAND USE & ENVTL L. 1 (2007).

Victor B. Flatt, *Adapting Laws for a Changing World: A Systemic Approach to Climate Change Adaptation*, 64 FLA. L. REV. 269 (2012).

J.B. Ruhl, *Climate Adaptation Law*, in GLOBAL CLIMATE CHANGE AND U.S. LAW (Michael B. Gerrard & Jody Freeman eds., 3rd ed, forthcoming 2022).

J.B. Ruhl, *Design Principles for Resilience and Adaptive Capacity in Legal Systems—with Applications to Climate Change Adaptation*, 89 N.C. L. REV. 1373 (2011).

J.B. Ruhl & James Salzman, *Climate Change Meets the Law of the Horse*, 62 DUKE L.J. 975 (2013).

Rob Verchick & Abby Hall, *Adapting to Climate Change While Planning for Disaster: Footholds, Rope Lines, and the Iowa Floods*, 2011 BYU L. REV. 2203 (2011).

Chapter 9

GEO-ENGINEERING

The mainstream conversation about climate change mitigation focuses on ways to reduce greenhouse gas emissions by changing the way that we use energy and by limiting deforestation. Emerging regulatory regimes incentivize measures such as using "clean coal," fuel switching—i.e., switching from coal to natural gas, ramping up sources of clean energy, or adopting energy efficiency and conservation measures. Much of the conversation centers on how to increase the percentage of electricity and transport fuel derived from renewable sources of energy or explores more traditional questions such as what role nuclear energy should play in a clean energy future. There are options for limiting the effects of climate change, however, that did not focus on reducing emissions.

This chapter will examine the growing area of geoengineering,[1] focusing on exploring the dual questions of what is geoengineering? And what role should the law play in regulating activities that fall under the broad umbrella of geoengineering? We will focus on developing a fuller understanding of the diverse range of tools and techniques that fall under the geoengineering label before exploring the relevant international governance debate. We will also look briefly at the domestic level to explore how geoengineering is, and is not, being regulated under existing federal law.

Geoengineering is a term that encompasses "planetary-scale, active interventions in the climate system to offset the build-up of greenhouse gases" (as opposed to methods of decreasing emissions).[2] To be classified as geoengineering, the scale of the project must usually be significant and the focus of the project must be on "deliberately alter[ing] the climate system."[3] Geoengineering is a broad term that can be used to cover numerous techniques and technologies designed to minimize or offset the effects of climate change. Most geoengineering proposals, however, fall under one of two broad categories: (1) efforts to remove greenhouse gas from the

[1] The terms geoengineering and climate engineering are sometimes used to refer to the same types of activities. Geoengineering is used here to encompass the broad range of activities that might also fall under the umbrella of climate engineering.

[2] David G. Victor, *On the Regulation of Geoengineering*, 24 OXFORD REV. ECON. POL'Y 322, 323 (2008).

[3] INTERGOVERNMENTAL PANEL ON CLIMATE CHANGE, CLIMATE 2014: SYNTHESIS REPORT, SUMMARY FOR POLICYMAKERS 12. FIFTH ASSESSMENT REPORT OF THE INTERGOVERNMENTAL PANEL ON CLIMATE CHANGE (2014).

atmosphere, and (2) solar radiation management, that is, efforts to reduce the amount of sunlight reaching the Earth.

While the debate surrounding geoengineering remains at the edges of popular and policy conversations about climate change, geoengineering research has expanded significantly in recent years giving rise to increasingly complex and contentious questions over the value, ethics, and governance of different proposed strategies. Because some proposed geoengineering technologies could potentially modify the climate of the Earth or a large portion of the Earth at an unusually fast pace, there is growing interest in what legal strategies to regulate the deployment of geoengineering projects.

I. Potential Methods of Geoengineering

As stated, there are two main categories into which geoengineering proposals fall. Each of these categories will be discussed in turn. It should be noted, however, that there is no broad agreement as to how to classify different responses to climate change, and there is at least some debate as to whether greater differentiation is needed both among the vastly different array of geoengineering technologies and between conventional mitigation techniques and geoengineering technologies.

A. Greenhouse Gas Removal

The first category of geoengineering includes a broad range of activities designed to remove greenhouse gases such as carbon dioxide (CO_2) from the atmosphere. Greenhouse gases can be removed from the atmosphere through efforts designed to modify terrestrial and oceanic ecosystems, or through the use of different technologies to remove greenhouse gases from the atmosphere. The IPCC defines carbon dioxide removal to include:

> anthropogenic activities removing CO_2 from the atmosphere and durably storing it in geological, terrestrial, or ocean reservoirs, or in products. It includes existing and potential anthropogenic enhancement of biological or geochemical sinks and direct air capture and storage, but excludes natural CO_2 uptake not directly caused by human activities.[4]

This category of activities is often referred to as carbon dioxide removal technologies (CDR), although the focus is on removing both

[4] INTERGOVERNMENTAL PANEL ON CLIMATE CHANGE, CLIMATE 2022: MITIGATION OF CLIMATE CHANGE, SUMMARY FOR POLICYMAKERS 40. SIXTH ASSESSMENT REPORT OF THE INTERGOVERNMENTAL PANEL ON CLIMATE CHANGE (2022).

carbon dioxide as well as other greenhouse gases from the atmosphere.

We start by discussing CDR technologies because this category includes approaches that are often seen as an extension of conventional mitigation strategies. In fact, in its Sixth Assessment Report, the IPCC treats certain CDR technologies (e.g., carbon dioxide capture and storage and afforestation) as mitigation tools and takes their deployment into account when considering different future mitigation scenarios. The IPCC even goes so far as to suggest that "the deployment of carbon dioxide removal (CDR) to counterbalance hard-to-abate residual emissions is unavoidable if net zero CO_2 or GHG emissions are to be achieved."[5] Increasingly, CDR activities are perceived to be politically and economically manageable way for large emitters, such as coal-fired power plants, to continue operating in the near term while also reducing their carbon footprint, as well as ways for governments to achieve aggressive net zero emissions targets.

For purposes of this discussion, we will further subdivide CDR technologies and explore approaches based on forestry/land use, oceanic system modification, and technology-based removal.

1. *Terrestrial Methods*

Terrestrial and ocean systems mitigate a large fraction of anthropogenic emissions by absorbing carbon emissions. As a result, many geoengineering techniques focus on maximizing the absorption capacity of both systems. The first method we will discuss focuses on maximizing the carbon absorption capacity of terrestrial systems.

Terrestrial methods for capturing carbon, including afforestation and reforestation, tend to be among "the least controversial and most politically acceptable geoengineering technique[s]."[6] Alongside fossil-fuel use in the electricity and transport sector, existing patterns of deforestation are one of the prime contributors to global carbon emissions. The IPCC estimates that agriculture, forestry, and other land use sector (AFOLU)

---

[5] INTERGOVERNMENTAL PANEL ON CLIMATE CHANGE, CLIMATE CHANGE 2022: MITIGATION OF CLIMATE CHANGE, SUMMARY FOR POLICYMAKERS 40. SIXTH ASSESSMENT REPORT OF THE INTERGOVERNMENTAL PANEL ON CLIMATE CHANGE (2022). *See also* MYLES ALLEN ET AL., INTERGOVERNMENTAL PANEL ON CLIMATE CHANGE, GLOBAL WARMING OF 1.5°C: SUMMARY FOR POLICYMAKERS 17 (Valérie Masson Delmotte et al. eds., 2018), https://www.ipcc.ch/sr15/chapter/spm/ [https://perma.cc/YFF8-SH6E] ("All pathways that limit global warming to 1.5C with limited or no overshoot project the use of carbon dioxide removal (CDR) on the order of 100–1000 GtCO2 over the 21st century. CDR would be used to compensate for residual emissions and, in most cases, achieve net negative emissions to return global warming to 1.5C following a peak.")

[6] Karen N. Scott, *International Law in the Anthropocene: Responding to the Geoengineering Challenge*, 34 Mich. J. Int'l L. 309, 322 (2013).

activities account for roughly a quarter of net anthropogenic GHG emissions, with much of this attributable to deforestation and agricultural emissions. As a result, efforts to curb existing patterns of deforestation, to reforest cleared areas, and to establish forest in new areas (primarily old agricultural land), provide important opportunities to combat climate change activities. Bent Sohngen, for example, estimates that forestry-related mitigation projects (including afforestation, reduced deforestation, and forest management) can contribute upwards of 30% towards global abatement efforts, while also reducing carbon prices up to 50%.[7] Forestry-related mitigation efforts, thus, represent significant, cost-effective measures for reducing greenhouse gas emissions.

Giving further credibility to this geoengineering subfield, both the Kyoto Protocol and the Paris Agreement specifically recognize the role that forestry-related activities can play in addressing climate change. The Kyoto Protocol directly recognizes that "net changes in greenhouse gas emissions by sources and removals by sinks resulting from direct human-induced land-use change and forestry activities, limited to afforestation, reforestation and deforestation"[8] could be used to meet emission reduction commitments made under the Protocol. Subsequent decisions of the Conference of the Parties created additional incentives, guidelines, and a work program for reducing emissions through forestry-related activities.

The Paris Agreement goes even further towards embracing forestry-related mitigations measures.[9] As one commentator notes, these forms of "soft-geo-engineering, through afforestation, reforestation, reduced deforestation and enhancement, [are] clearly encouraged by the Agreement."[10] The Paris Agreement embraces a broader source of forest-related activities, recognizes the importance of financing for these activities, and encourages such support from both public and private sources.

Forestry-related mitigations efforts are coordinated under the UN REDD+ program.[11] The parties to the UNFCCC created REDD+

---

7 Brent Sohngen, *An Analysis of Forestry Carbon Sequestration as a Response to Climate Change* (2008), Copenhagen Consensus on Climate, http://aede.osu.edu/sites/aede/files/publication_files/Analysis%20of%20Forestry%20Carbon.pdf.

8 Kyoto Protocol to the United Nations Framework Convention on Climate Change, Dec. 10, 1997, UN Doc. FCCC/CP/1997/7/Add.2, Dec. 10, 1997, *reprinted in* 37 I.L.M. 22 (1998) [hereinafter Kyoto Protocol]. The Kyoto Protocol opened for signature March 16, 1998, and entered into force February 16, 2005.

9 Paris Agreement, Preamble, Dec. 12, 2015, U.N. Doc. FCCC/CP/2015/L.9/Rev.1, art. 5 (entered into force Nov. 4, 2016).

10 Jorge E. Viñuales, *The Paris Climate Agreement: An Initial Examination*, in C-EENRG WORKING PAPERS, NO. 6, 15 December 2015.

11 United Nations Climate Change, *What is REDD+?*, https://unfccc.int/topics/land-use/workstreams/redd/what-is-redd. REDD stands for reducing emission from forest degradation in developing countries.

in 2013 as a framework for guiding forestry-related mitigation and sustainability efforts. REDD+ is a voluntary framework that establishes methodological and financing guidance for any party that is interested in engaging in forestry-related activities designed to achieve emissions reductions. As David Takacs describes:

> In a REDD+ project or program, an individual landowner, local community, private developer, or government entity reforests degraded land or pledges to preserve a forest that would otherwise be felled. They may then sell the sequestered carbon for a contracted period of time to entities that want to offset their GHG emission[s]. REDD+ happens on a project-by-project basis, where a developer contracts with landowners to preserve or reforest land, and sells the stored carbon. Alternatively, nations, states, or provinces implement REDD+ on a broader, "jurisdictional" scale, i.e., they use REDD+ funds to reduce deforestation or promote reforestation in a broad geographic area, resulting in greater stored carbon than would have occurred absent the funding.[12]

The Paris Agreement formally recognizes REDD+ and encourages parties to the Agreement to engage in REDD+ activities to achieve the goals of the Paris Agreement.

Terrestrial methods for capturing carbon are becoming widely embraced as generally acceptable methods for limiting climate change. These methods are not without controversy, including pervasive concerns about the displacement of peoples living in forested areas, leakage of land-use activities from one place to another, and proper accounting for emissions reductions. The positive benefits of investing in these activities, however, means that the suite of terrestrial activities viewed as mainstream forms of mitigation continues to grow. Although many of these activities continue to be lumped under the very broad heading of geoengineering, because they are frequently treated as mainstream forms of mitigation, there are calls for greater differentiation between efforts such as these and other more controversial forms of climate modification.

2. *Oceanic System Modification*

The second category of CDR activities includes technologies to modify the ocean system to increase carbon absorption (mari-engineering). As Eagle and Sumaila point out, these engineering

---

[12] David Takacs, *Protecting Your Environment, Exacerbating Justice: Avoiding "Mandate Havens"*, 24 DUKE ENVTL. L. & POL'Y F. 315, 342 (2014).

efforts "represent[s] an ocean use never before undertaken on a large scale and thus not currently regulated."[13]

Oceans are the Earth's most important carbon sink. Although estimates vary, recent research indicates that approximately 26 percent of all the carbon released as CO_2 from fossil fuel burning, cement manufacture, and land-use changes between 2002–2011 was absorbed by the oceans, meaning that over a quarter of all carbon is absorbed by our oceans.[14] There is a risk, however, that climate change could reduce the capacity of the oceans to absorb carbon by approximately 9 percent.[15] As a result, maintaining and expanding this absorption capacity is of upmost importance to ongoing efforts to limit climate change and extensive research has examined how to enhance natural oceanic pumps that drawdown carbon from the atmosphere into the depths of the oceans.

The most discussed method for modifying the ocean system is through a process known as ocean fertilization. Phytoplankton, which are the dominant form of plant in the open ocean, serve as a biological pump for drawing down carbon from the surface into the deep seas. Using sunlight for energy, phytoplankton "convert dissolved inorganic carbon . . . in seawater into organic matter through photosynthesis, driving global marine food webs and prompting the 'drawdown' of additional carbon dioxide from the atmosphere."[16] Although phytoplankton are pervasive throughout the oceans, they appear in greater numbers in certain parts of the oceans, while other parts of the oceans (e.g., the Southern Ocean) have significantly lower phytoplankton populations, resulting in a reduced carbon drawdown potential in these areas. Phytoplankton populations are lower in certain parts of the oceans primarily because of low levels of essential nutrients, including iron, as well as factors such as strong turbulence, lack of light, and competition.

Recognizing this phenomenon, John Martin and colleagues hypothesized that the availability of micronutrients, such as iron,

---

[13] Josh Eagle & U. Rashid Sumaila, *Climate, Oceans, the Law of Special and General Adaptation,* in THE OXFORD HANDBOOK OF INTERNATIONAL CLIMATE CHANGE LAW (Cinnamon Carlarne, Kevin R. Gray & Richard Tarasofsky eds. (2016)).

[14] C. Le Quéré et al., *The Global Carbon Budget 1959–2011,* 5 EARTH SYST. SCI. DATA DISCUSS. 1107–1157 (2012), http://www.earth-syst-sci-data-discuss.net/5/1107/2012/essdd-5-1107-2012.pdf. *See also* Andrew J. Watson et al., *Revised Estimates of Ocean-Atmospheric CO2 Flux are Consistent with Ocean Carbon Inventory,* 11 NATURE COMMUNICATIONS 4422 (2020).

[15] Andy J. Ridgwell, Mark A. Maslin, & Andrew J. Watson, *Reduced Effectiveness of Terrestrial Carbon Sequestration Due to an Antagonistic Response of Ocean Productivity,* 29 GEOPHYSICAL RESEARCH LETTERS 1095 (2002).

[16] Secretariat of the Convention on Biological Diversity, *Scientific Synthesis of the Impacts of Ocean Fertilization on Marine Biodiversity,* CBD TECHNICAL SERIES NO. 45, 15, https://www.cbd.int/doc/publications/cbd-ts-45-en.pdf.

function as a limiting factor in phytoplankton photosynthesis.[17] Further research suggests that the availability of iron is a key factor influencing the associated uptake of carbon over large areas of the ocean, particularly in parts of the oceans, including the Southern Ocean, that have high potential to function as more effective carbon sinks. As a result of this research, scientists began to study ways to increase carbon uptake in these areas by facilitating phytoplankton bloom through the addition of iron dust to the surface of the ocean. This technique became known as ocean iron fertilization.

Since John Martin's famous statement in 1991, when he declared, "[g]ive me a half a tanker of iron and I will give you another ice age"[18], ocean iron fertilization research and experimentation has been one of the highest-profile forms of geoengineering and, as will be discussed later in this chapter, has been the subject of significant legal and political attention at the international level.

To date, researchers have engaged in a handful of small-scale iron fertilization experiments. While most of these experiments have led to an increase in the number of phytoplankton, the effectiveness of fertilization as a method for limiting climate change remains unproven due to incomplete understanding of other factors limiting phytoplankton growth, lack of evidence as to whether the CO_2 absorbed at the surface has been drawn down and sequestered in the deep ocean, and concerns about the effects of these experiments on the ocean's biological and chemical processes. In key part, "possible effects include increased ocean acidification, the disruption of marine ecosystems, eutrophication and anoxia, the creation of toxic harmful algal blooms, the generation of an increase in the emission of other greenhouse gases such as nitrous oxide, and a decrease in the effectiveness of the Southern Ocean methyl bromide sink leading to a delay in the recovery of the ozone layer."[19] As Wallace et al. caution:

> Large-scale fertilization could have unintended (and difficult to predict) impacts not only locally, e.g. risk of toxic algal blooms, but also far removed in space and time. Impact assessments need to include the possibility of such 'far-field' effects on biological productivity, sub-surface oxygen levels, biogas production and ocean acidification.[20]

As a result of the many unknowns associated with iron fertilization and concerns about the impact on biodiversity, several

---

[17] See John H. Martin et al., *Testing the Iron Hypothesis in Ecosystems of the Equatorial Pacific Ocean*, 371 NATURE, Vol. 123 (1994).

[18] CBD TECHNICAL SERIES NO. 45, *supra* note 16.

[19] Scott, *supra* note 6, at 324–25.

[20] Doug Wallace et al., *Ocean Fertilization: A Scientific Summary for Policymakers* 1, INTERGOVERNMENTAL OCEANOGRAPHIC COMMISSION (2015), http://unesdoc.unesco.org/images/0019/001906/190674e.pdf.

international bodies have banned or urged caution with respect to future iron fertilization experiments.

While ocean fertilization has received the most attention in international forums, researchers have been studying other ways to increase the oceans' uptake of carbon, including use of other fertilization methods (e.g., volcanic ash); using vertical ocean pipes to move seawater from the depths of the ocean to the surface to improve cooling and thus uptake; taking advantage of the ocean's solubility pump (a naturally occurring process whereby dissolved inorganic carbon is transferred from the ocean's surface to its interior via a down-welling current) by directly injecting captured carbon into these down-welling currents; and, increasing the alkalinity of the ocean through the addition of limestone powder or soda ash, thus improving the oceans absorption capacity and limiting the negative effects of ocean acidification. Most of these ideas are still in the research phase or are considered unviable either due to economic infeasibility or concerns about negative, unintended consequences.

As the global climate system warms, scientists have been increasingly concerned about the effects of climate change on the oceans, including concerns about thermal expansion and ocean acidification. The majority of ocean-based geoengineering proposals do not seek to remedy these ocean-specific challenges, instead focusing on reducing total atmospheric concentrations of carbon dioxide. The narrow focus of these proposals creates additional concerns for scientists focused on the health of the ocean system and the plants, animals, and humans who depend upon it.

3. *Technology-Based Removal*

The third and final category of CDR involves efforts to prevent CO_2 from being emitted to the atmosphere by capturing, compressing, transporting, and storing it. The most common technique is referred to as carbon capture and storage (CCS).

Carbon can be captured before or after combustion. There are three primary approaches to CCS. These include:

- As a pure or near-pure CO_2 stream, either from an existing industrial process or by reengineering a process to generate such a stream (e.g., oxyfueling power-generation plant, precombustion fuel gasification).
- Concentration of the discharge from an industrial process into a pure or near-pure CO_2 stream (e.g., post-combustion separation from power plant or cement plant flue gases).

- Direct air capture into a pure CO_2 stream or into a chemically stable end product (e.g., mineralization of steel slag).[21]

Carbon capture is most often associated with efforts to capture the CO_2 produced by traditional fossil fuel powered electric generating units (EGUs), particularly from coal-fired units, where the carbon is captured during a post-combustion process. There is increasing interest, however, in technologies that would remove CO_2 from the atmosphere.

Regardless of how CO_2 is captured, it still has to be safely and permanently stored or "sequestered" in order to provide an effective tool for limiting climate change. Sequestration is a "process whereby CO_2 is captured from large stationary point-sources (e.g., coal-fired power plants, ethanol refineries, and cement manufacturers), compressed, and transported by pipeline to locations where it is injected deep into the subsurface for isolation from the atmosphere."[22] The captured carbon can be stored in various forms and locations, including in the deep ocean, depleted oil and gas wells, or deep saline aquifers; and as manufactured mineralized carbonate rock, or as naturally mineralized carbonate by injection into basalt reservoirs. The goal is to permanently remove the CO_2 from the atmosphere, thus allowing energy to be produced with a much smaller carbon footprint. While approaches vary, it is estimated that the amount of emissions reduced using CCS is approximately 80–100%, with the cost of coal-fired electricity generation increasing by at least the same amount.

There is an existing body of state case law that explores the injection of CO_2 into underground spaces to enhance oil and gas recovery, but the practice of injecting large amounts of CO_2 into underground spaces for geologic sequestration is relatively new and the research, technology, and applicable federal legal regime are all still in nascent stages. There are also significant state law issues regarding the ownership of sequestration capacity far underground and potential tort liability for releases. Despite being in the early stages of development, in the United States, CCS is increasingly looked to as a way to continue using existing energy infrastructure while simultaneously making headway in efforts to limit climate change. President Obama's Clean Power Plan, for example, would have required any new coal power plant built in the United States to emit no more than 1,400lbs of CO_2 per megawatt hour. Based on

---

[21] STEPHEN A. RACKLEY, CARBON CAPTURE AND STORAGE 19 (2010).

[22] Jeffrey M. Bielicki et al., *An Examination of Geologic Carbon Sequestration Policies in the Context of Leakage Potential*, 37 INT. J. GREENH. GAS CONTROL 61, 61 (2015) (citing the IPCC, 2005. IPCC SPECIAL REPORT ON CARBON DIOXIDE CAPTURE AND STORAGE (Bert Metz et al., eds. 2005)).

existing technology at the time, this would essentially have necessitated that any new power plant makes some use of CCS to reach this standard.

However, the viability of using CCS technology in this way remains untested, with many critics questioning whether CCS constitutes an "adequately demonstrated" technology,[23] as required under the U.S. Clean Air Act (CAA). As of 2022, there were only twenty-seven commercial CCS projects in the world, about half in the United States; and all but one use the captured CO_2 to enhance oil extraction.[24] The costs associated with CCS are high, whether a facility is investing in retrofitting or building from the ground up, and "there is broad agreement that costs for CCS would need to decrease before the technologies could be widely deployed across the nation."[25] While the incremental cost of CCS varies considerably depending on the nature of the project, current carbon capture costs are estimated at $43–$65 per ton CO_2 captured.[26] CCS, thus, would only be profitable if there was a high price on carbon.

The heavy and variable costs associated with using CCS mean that rapid uptake of this technology is dependent on declining costs, subsidization, or regulatory limitations and incentives such as those that the Biden Administration is currently pursuing.

The range of geoengineering technologies falling under the umbrella of CDR, including land-use modification, ocean system modification, and technology-based carbon capture techniques, is extremely broad. By and large, however, the most dominant forms of

---

[23] For purposes of the Clean Air Act, an "adequately demonstrated" system of emission reduction is "one which has been shown to be reasonably reliable, reasonably efficient, and which can reasonably be expected to serve the interests of pollution control without becoming exorbitantly costly." Essex Chem. Corp. v. Ruckelshaus, 486 F.2d 427, 433 (D.C. Cir. 1973), *cert. denied*, 416 U.S. 969 (1974).

[24] Jonathan M. Moch, William Xue, John Holdren, *Carbon Capture, Utilization, and Storage: Technologies and Costs in the U.S. Context*, Harvard Kennedy School, Belfer Center for Science and International Affairs, Policy Brief (Jan. 2022), https://www.belfercenter.org/publication/carbon-capture-utilization-and-storage-technologies-and-costs-us-context. *See also* Congressional Research Service, *Carbon Capture and Sequestration (CCS) in the United States* 13 (Oct. 18, 2021), https://sgp.fas.org/crs/misc/R44902.pdf.

[25] *Id.*

[26] Congressional Research Service, *Carbon Capture and Sequestration (CCS) in the United States* 13 (Oct. 18, 2021), https://sgp.fas.org/crs/misc/R44902.pdf. *See also* William J. Schmelz, Gal Hochman & Kenneth G. Miller, *Total Cost of Carbon Capture and Storage Implemented at a Regional Scale: Northeastern and Midwestern United States*, The Royal Society (Aug. 14, 2020), https://royalsocietypublishing.org/doi/10.1098/rsfs.2019.0065 ("suggest[ing] coal-sourced CO_2 emissions can be stored in [the northeastern and midwestern United States] at a cost of $52–$60 ton, whereas the cost to store emission from natural-gas-fired plants ranges from approximately $80 to $90."); Adam Baylin-Stern & Niels Berghout, *Is Carbon Capture too Expensive?*, IEA (Feb. 27, 2021), https://www.iea.org/commentaries/is-carbon-capture-too-expensive.

these technologies have received considerably more legal and political attention than the second category of geoengineering proposals that we will consider, which fall under the heading of solar radiation management (SRM).

B. Solar Radiation Management

In contrast to CDR techniques, which focus on preventing greenhouse gases from reaching the atmosphere, SRM techniques focus on modifying the albedo of the planet. Albedo refers to the fraction of solar radiation reflected by a surface or object. For example, white, snow-covered surfaces have a high albedo, with the white color helping the surface to reflect higher levels of solar radiation than would a dark, vegetation-covered surface or the ocean. The level of albedo is referred to as high (e.g., snow) to low (e.g., dark vegetation).[27]

The goal of most proposed SRM techniques is to lower the temperature of the Earth's surface by decreasing the amount of energy that the Earth absorbs by increasing the albedo, or reflectivity of the planet. The general idea is that the more we can deflect light away from the surface of the Earth and back into the atmosphere, the cooler we can keep the planet. As the IPCC explains, "SRM contrasts with climate change mitigation activities, such as emissions reductions and carbon dioxide removal, as it introduces a 'mask' to the climate change problem by altering the Earth's radiation budget, rather than addressing the root cause of the problem, which is the increase in greenhouse gas concentrations in the atmosphere."[28] These techniques, thus, are not traditional mitigation tools because they do not focus on human anthropogenic forcing; that is, they do not seek to change land-use or greenhouse gas emission patterns.

Thus, instead of mitigating greenhouse gases, SRM focus on finding ways to offset the negative effects of climate change separate to efforts to limit climate change through changes in emissions scenarios. Given that is has been estimated that "a deflection of approximately 1.8% of solar radiation would offset the global mean temperature effects of a doubling of atmospheric concentrations of CO_2,"[29] SRM is often presented as an important option to explore in case mitigation efforts fail and humans need some type of backstop

---

[27] INTERGOVERNMENTAL PANEL ON CLIMATE CHANGE, *Glossary*, in CLIMATE CHANGE 2007: WORKING GROUP I: THE PHYSICAL SCIENCE BASIS 941 (Alphonsus P. M. Baede ed., 2007), https://www.ipcc.ch/report/ar4/wg1/.

[28] INTERGOVERNMENTAL PANEL ON CLIMATE CHANGE, CLIMATE CHANGE 2022: MITIGATION OF CLIMATE CHANGE 2017. SIXTH ASSESSMENT REPORT OF THE INTERGOVERNMENTAL PANEL ON CLIMATE CHANGE (2022).

[29] Paul J. Crutzen, *Albedo Enhancement by Stratospheric Sulfur Injections: A Contribution to Resolve a Policy Dilemma?*, 77 CLIMATIC CHANGE 211, 216 (2006).

method for relieving the negative effects of climate change. It is important to understand, however, that these techniques neither seek to change the short or long-term emissions profile nor to address other critical climate-related effects such as ocean acidification.

In common with CDR, there are a wide variety of techniques that fall under the umbrella of SRM. These include relatively mundane proposal such as enhancing the albedo of urban environments by, for example, increasing rooftop albedo by using lighter colors and more reflective materials, to increasingly complex and controversial proposals ranging from suggestions to enhance the albedo of natural environments by using heat reflecting sheets to cover arid areas, to switching to natural or bioengineered grasses, shrubs, and crops that are lighter in color and more reflective than existing flora, to placing mirrors in space or injecting aerosols into the stratosphere. Burns describes some of the boldest SRM proposals:

> Several proposals involve placing reflectors in near-Earth orbits, including placement of 55,000 mirrors in random orbits, or the creation of a ring of dust particles guided by satellites at altitudes of approximately 1,200 to 2,400 miles. An alternative approach could be to establish a "cloud of spacecraft" with reflectors in a stationary orbit near the Inner Lagrange point (L1), a gravitationally stable point between Earth and the sun.[30]

Of the various SRM proposals, the category that has been the subject of the most research and debate is the potential introduction of aerosols into the stratosphere to cool the climate. It has long been understood that aerosols—small particles that reflect sunlight—can have a cooling influence on the climate, with proposals to use aerosols in this way dating back to the 1970s. Aerosols are produced naturally through, for example, volcanic eruptions, as well as through man-made processes, such as the combustion of coal. Most recent proposals focus on introducing chemical precursors for aerosols, such as sulfur dioxide, into the stratosphere, which is the layer of the atmosphere where rainfall and most conventional "weather" occurs. Once introduced into the stratosphere, these chemical precursors transform into aerosols and can induce a cooling effect. Aerosols are very effective reflectors of sunlight, so the mass needed to achieve a significant amount of reflection and, thus, cooling is relatively small. As a recent National Academy of Science (NAS) report concluded:

> To put the required increase in albedo into perspective, the 1991 Pinatubo eruption, which is estimated to have been the largest eruption since Krakatau in 1883, led to a

[30] William C.G. Burns, *Geoengineering the Climate: An Overview of Solar Radiation Management Options,* 46 TULSA L.R. 283, 295 (2012).

radiative forcing of approximately −3 W/m2 within a month following the eruption, decreasing to nearly zero over the subsequent 2 years and causing the average surface air temperature to cool an estimated 0.3°C over a period of 3 years.[31]

Thus, the strength of stratospheric sulfate aerosols is that a small mass can produce a significant amount of cooling. The risks of using this approach, however, are many and poorly understood. Aerosols are among the most difficult components of the climate system to model and monitor and are, thus, one of the most poorly understood parts of the climate system. As a recent report by the National Academy of Sciences notes, "currently, observational capabilities lack the capacity to monitor the evolution of an albedo modification deployment (e.g., the fate of the aerosols and secondary chemical reactions), its effect on albedo, or its environmental effects on climate or other important Earth systems."[32] The range of unknown and, currently, undiscoverable dangers surrounding the use of stratospheric sulfate aerosols make these proposals extremely risky.

This is especially true because their use, in common with other SRM methods, would temporarily mask high atmospheric concentration of greenhouse gases and resulting warming. Thus, if aerosols are used and then their use is ceased, the Earth would rapidly confront significantly higher temperatures than experienced prior to the deployment of the aerosols. The risks are also high because of the potential for serious distributive justice concerns. If, for example, an actor engages in an aerosol injection project that results in cooling in the Northern Hemisphere, there is the potential for negative effects (e.g., changing rain patterns and drought) in the Southern Hemisphere.[33] There are also concerns about intergenerational justice, that is, the impacts that aerosol-based experimentation might have on future generations by, for example, creating scenarios where future generations are locked into certain climate response pathways by the decisions of current generations.

Concerns about SRM techniques have constrained public debate and policy making around this topic.[34] As the IPCC suggests, "solar

---

[31] National Academy of Sciences, Committee on Geoengineering Climate, *Climate Intervention: Reflecting Sunlight to Cool the Earth* 35 (2015) [hereinafter National Academy of Sciences].

[32] *Id.* at 8.

[33] *See, e.g.*, Jim M. Haywood et al., *Asymmetric Forcing from Stratospheric Aerosols Impacts Sahelian Rainfall*, NATURE CLIMATE CHANGE (March 31, 2013).

[34] The IPCC describes some of these concerns and knowledge gaps:

> Solar radiation modification (SRM) measures are not included in any of the available assessed pathways. Although some SRM measures may be theoretically effective in reducing an overshoot, they face large uncertainties

radiation modification approaches, if they were to be implemented, introduce a widespread range of new risks to people and ecosystems, which are not well understood", and "large uncertainties and knowledge gaps are associated with the potential of solar radiation modification approaches to reduce climate change risks."[35] However, because tools such as stratospheric sulfate aerosol injection potentially provide methods for limiting the negative effects of climate change relatively quickly through potentially modest expenditures of effort and capital, there is an active debate over whether the failure to research and experiment with these strategies limits our ability to respond should the negative effects of climate change arise more quickly or dramatically than predicted.

II. Geoengineering Governance

Although the past decade has seen a dramatic increase in the amount of research and public debate surrounding the topic, geoengineering is not a new phenomenon. For years, humans have attempted to directly affect the climate system in a multitude of ways. Most of these early efforts, however, were small-scale and localized. In the context of global climate change, the scale and stakes of geoengineering are much grander. As global climate change intensifies and as international efforts to respond to climate change continue to struggle to motivate aggressive mitigation strategies, scientists and policymakers worldwide increasingly seek innovative and, at times, unilateral ways to limit the negative consequences of climate change. As a result, over the past decade, many of the geoengineering techniques described above have moved from the periphery into mainstream research and policy forums.

The recent attention paid to geoengineering both in the popular media, and in high-level political and academic forums is attributable to a number of factors, including growing awareness of the risks associated with climate change and continuing gaps between mitigation needs and commitments. In particular, increased understanding about the pervasive impacts of climate change, coupled with concern about tipping points and non-linear changes in the climate system have amplified fears about abrupt climate change and prompted scientists and policymakers to look for more drastic

---

> and knowledge gaps as well as substantial risks and institutional and social constraints to deployment related to governance, ethics, and impacts on sustainable development. They also do not mitigate ocean acidification.

MYLES ALLEN ET AL., INTERGOVERNMENTAL PANEL ON CLIMATE CHANGE, GLOBAL WARMING OF 1.5°C: SUMMARY FOR POLICYMAKERS 12–13 (Valérie Masson Delmotte et al. eds., 2018), https://www.ipcc.ch/sr15/chapter/spm/ [https://perma.cc/YFF8-SH6E].

[35] INTERGOVERNMENTAL PANEL ON CLIMATE CHANGE, CLIMATE CHANGE 2022: MITIGATION OF CLIMATE CHANGE, SUMMARY FOR POLICYMAKERS 20, SIXTH ASSESSMENT REPORT OF THE INTERGOVERNMENTAL PANEL ON CLIMATE CHANGE (2022).

ways to limit the negative effects of climate change should political efforts fail—a Plan B.

Recognizing the limits of traditional policy options, even when considering the relatively positive outcome at the 2015 Paris and 2021 Glasgow COPs, efforts have grown to facilitate a more open conversation about the pros and cons of different geoengineering techniques, discourage clandestine and unilateral actions, and encourage cooperation around research and governance. These efforts also emphasize four other aspects of geoengineering: (1) the possible economic benefits of geoengineering, (2) the potential for rapid realization of benefits, (3) the reality that small-scale geoengineering experiments are already taking place, and (4) the fear of significant or irreversible unilateral actions being taken in the near future absent a shared system of norms or governance.

First, in comparison to the costs of traditional greenhouse gas mitigation strategies, the economics of geoengineering have been referred to as "incredible."[36] Although the costs of different geoengineering techniques vary widely, there is a general perception that, in contrast to global mitigation options, certain geoengineering tools, particularly certain SRM techniques, can be used to halt global warming at a fraction of the cost. As described by one commentator, "while [greenhouse gas] mitigation policies involve a substantial number of actors undertaking expensive policies that require highly decentralized implementation, geoengineering involves a small number of actors undertaking highly centralized, relatively inexpensive actions."[37] The ability of single states—or even extremely wealthy individuals—to fund and deploy geoengineering technologies stands in stark contrast to the administrative and financial complexity associated with globally or nationally organized, economy-wide efforts to abate greenhouse gas emissions. In this way, certain geoengineering tools offer more straightforward, but less transparent and politically accountable pathways for attempting to limit the negative effects of climate change.

Similarly, in contrast to the complex global politics of climate change, involving historically and politically fraught questions over which countries should bear the burden of limiting greenhouse gas emissions and to what degree, certain geoengineering approaches would not involve analogous competitive advantage/disadvantage calculations. For example, while one country may not find it politically or economically optimal to aggressively abate greenhouse

---

[36] Scott Barrett, *The Incredible Economics of Geoengineering*, 39 ENVTL. & RES. ECON. 45, 45–46 (2008).

[37] William Daniel Davis, *What Does "Green" Mean?: Anthropogenic Climate Change, Geoengineering, and International Environmental Law*, 43 GA. L. REV. 901, 925–26 (2009).

gas emissions if another country abstains from doing so, that country still might find it both politically and economically viable to deploy geoengineering technologies (e.g., a sulfate aerosol program) to try to offset climate change regardless of the acts or omissions of its global counterparts. And, while the estimated costs of many geoengineering techniques are often vastly understated due to unaccounted for associated and indirect costs related to the possibility of unforeseen climatic responses, localized harms, political conflict, and intergenerational effects, the ability to approximate the direct costs of deploying a specific technology to halt climate change offers an overly simply, but attractive alternative to the complexity that characterizes the economics of conventional policy choices.

Second, beyond the perceived economic benefits of certain approaches, geoengineering also offers the possibility of rapid positive change. Many conventional mitigation strategies focus on complex processes such as overhauling transportation and energy infrastructure. These changes occur slowly and incrementally, and the benefits of these changes will often not be fully felt for generations due to the long lifespan of some greenhouse gases in the atmosphere. In contrast, certain geoengineering opportunities offer at least the possibility of near-term benefits, which may eventually make them more politically palatable than strategies that ask constituents to make sacrifices for benefits that they themselves may never see. If, for example, a country opts to deploy an albedo modification technology that leads to near-term reductions in global temperatures, the political leaders may reap political goodwill along with the benefits of temporary, regional climate stabilization regardless of whether it is undertaking long-term, globally beneficial efforts to abate greenhouse gas emissions. While there are critical uncertainties related to the ability of different techniques to halt long-term climate change, the promise of visible results, in real time, at seemingly modest cost, may lead to increased interest in geoengineering options down the road.

The third reason that geoengineering is receiving increased political attention is because, as geoengineering research grows globally, so too does small-scale experimentation. While estimates vary and there is a dearth of reliable data, as previously mentioned, there are at least twenty-seven commercial-scale CCS facilities at industrial plants around the world, a handful of ocean iron fertilization experiments have taken place, there are widespread land-use based efforts, and growing calls for controlled experimentation with stratospheric sulfate aerosols. In other words, this is a burgeoning field where experimentation is expected to grow. Given that the stakes of experimentation will only increase with time, there is an urgent need think preemptively about norms and governance.

Hand-in-hand with the growth in experimentation, as geoengineering research become more sophisticated and as small-scale experimentation begins to yield results, it becomes more and more likely that one or more actors might decide to engage in larger scale experimentation. It seems wise to craft a governance structure before matters reach that stage. This leads to concerns that a single, or a handful of actors could decide to take actions which could bring about serious, negative consequences, which is the fourth reason that the debate over geoengineering governance has become increasingly mainstream.

A. International Efforts to Govern Geoengineering

At the moment, in contrast to the collective body of norms and rules defining the field of global climate change law, there is no targeted international governance regime for geoengineering. The relative absence of shared norms or rules around geoengineering means that it may be possible for geoengineering decisions to be made unilaterally, or for public or private actors to engage in small-scale experimentation without notice or clear means of accountability. This is a cause of concern for some. This section explores some of the norms and rules that are emerging with respect to geoengineering at both the international and domestic levels before examining the question of what more is needed.

At the international level, as Scott notes, "[w]ith the exception of reforestation and afforestation and ocean fertilization for scientific research purposes, there are few legal instruments explicitly applicable to geoengineering; however, as an activity that creates a significant risk of serious harm to the environment, geoengineering is subject to the obligations and principles of international environmental law more generally."[38] While this chapter will not explore the principles of international environmental law in depth, it is worth noting that there are a set of principles that, to varying extents of consensus and clarity define the parameters of the field of international environmental law and state obligations in international law. Although there have not yet been any direct legal efforts to apply these principles to geoengineering practices in such a way as to constrain behavior or determine accountability, at least one commentator suggests that these principles are "sufficiently extensive in their scope and nature of application to constrain at least some of the proposed techniques of geoengineering."[39] For example,

[38] Scott, *supra* note 6, at 330.

[39] *Id.*

questions of State Responsibility may arise if geoengineering techniques cause transboundary harm, while existing and emerging principles of international law, such as, the prevention of harm, the precautionary principle, the principle of permanent sovereignty over natural resources, and the interrelated principles of sustainable development, and intergeneration equity offer alternative ways to approach the regulation (ex ante) or remediation (ex post) of geoengineering activities. As of yet, however, this body of vaguely applicable provisions and principles offers—at best—only a starting point for developing a set of guiding norms for geoengineering governance. In common with the larger challenge of climate change, however, absent greater development, it is unlikely that these existing principles alone, or in aggregate, provide an adequate normative baseline, or behavior backstop for the complex array of activities that could be conducted under the heading of geoengineering.

In addition to the obligations and principles of international law, various international instruments include provisions that are relevant but not yet directly applicable to geoengineering. For example, the 1977 Environmental Modification Convention (ENMOD) prohibits climate manipulation for military or hostile use. With respect to the oceans and potential efforts to modify the oceanic system, the United Nations Convention on the Law of the Sea (UNCLOS) regulates pollution of the seas, the Convention on the Prevention of Marine Pollution by Dumping of Wastes and Other Matter (London Convention) prohibits the dumping of certain hazardous materials and requires a prior permit for the dumping of a number of other wastes and identified materials. Concerning the prospect that a variety of geoengineering practices, including the use of stratospheric sulfate aerosols, could cause transboundary harm, the Convention on Environmental Impact Assessment in a Transboundary Context (Espoo Convention), alongside the International Law Commission's Draft Articles on the subject of the prevention of significant transboundary harm from hazardous activities and the recent decision of the International Court of Justice in *Pulp Mills on the River Uruguay*,[40] provide the parameters for an emerging body of international law that constrains domestic activities that cause transboundary harm and would be applicable should a geoengineering project result in transboundary harm, or

[40] Pulp Mills on the River Uruguay (Arg. v. Uru.), Judgment, 2010 I.C.J. 20, ¶¶ 178–80 (Apr. 20); *see also* Cymie R. Payne, Pulp Mills on the River Uruguay: *The International Court of Justice Recognizes Environmental Impact Assessment as a Duty Under International Law*, ASIL INSIGHT (Apr. 22, 2010), https://www.asil.org/insights/volume/14/issue/9/pulp-mills-river-uruguay-international-court-justice-recognizes (discussing the implications of the case for notification, consultation and broadened EIA requirements in regards to activities that pose a risk of transboundary harm).

pose the concrete potential to do so. In addition, the Convention on Biological Diversity provides for the conservation of biological diversity and, thus, applies to any practice, such as ocean fertilization, that might pose a risk to marine biodiversity.

Moving from the general to the specific, the constituent bodies for at least two multilateral international institutions, the London Convention and the Convention on Biological Diversity (CBD), have taken discrete steps towards the governance of one specific geoengineering technology—ocean fertilization.

In 2007, the Scientific Group of the London Convention and the London Protocol released a *Statement of Concern Regarding Iron Fertilization of the Oceans to Sequester CO2*. In the statement, the Group acknowledged commercial interest in the technology of ocean-iron fertilization and cautioned that existing knowledge about the practice is insufficient to justify large-scale experimentation. Citing the findings of the IPCC that, while ocean-iron fertilization may offer "a potential strategy for removing carbon dioxide from the atmosphere," it "remains largely speculative, and many of the environmental side effects have yet to be assessed," the Group "note[d] with concern the potential for large-scale ocean iron fertilization to have negative impacts on the marine environment and human health."[41] In light of these concerns, the Group elaborated a set of detailed factors that should be considered in evaluating whether any ocean-iron fertilization projects should move forward.

In the following years, the parties to the Convention and the Protocol endorsed the Statement of Concern developed by the Scientific Group, agreed that questions of ocean-iron fertilization fell within the remit of the Convention and the Protocol, in view of their common objective of protecting and preserving the marine environment. It also recognized that each State maintained the right to consider ocean iron fertilization proposals on a case-by-case basis while also urging States to "use the utmost caution" when considering such proposals. The issue of iron fertilization continues to be studied by a special working group formed under the Convention and the Protocol. From time to time, the parties have considered proposals to amend the London Protocol to regulate ocean fertilization and other marine geo-engineering activities. At this

---

[41] Int'l Mar. Org. [IMO], *Statement of Concern Regarding Iron Fertilization of the Oceans to Sequester CO2*, ¶ 1 IMO Ref. T5/5.01, LC-LP.1/Circ. 14 (July 13, 2007), https://www.whoi.edu/cms/files/London_Convention_statement_24743_29324.pdf; *see also* Ninth Meeting of the Conference of the Parties to Convention on Biological Diversity, Bonn, Germany, May 19–30, 2008, Decision IX/16: Biodiversity and Climate Change, § C, UNEP/CBD/COP/DEC/IX/16 (Oct. 9, 2008), http://www.cbd.int/doc/decisions/cop-09/cop-09-dec-16-en.pdf.

time, however, the Parties to the Convention and the Protocol have not yet taken direct action to regulate the practice of ocean-iron fertilization. Nevertheless, the London Convention and Protocol were among the first international institutions to deliberate around the need to regulate geoengineering.

Following in the footsteps of the London Convention, in 2008, the Conference of the Parties (COP) to the CBD issued a decision welcoming the decision by the London Convention and Protocol to urge caution with respect to proposals for large-scale ocean fertilization. The COP then incorporated the question of iron fertilization into its working agenda and issued a decision that:

> *[R]equests* Parties and *urges* other Governments, in accordance with the precautionary approach, to ensure that ocean fertilization activities do not take place until there is an adequate scientific basis on which to justify such activities, including assessing associated risks, and a global, transparent and effective control and regulatory mechanism is in place for these activities; with the exception of small scale scientific research studies within coastal waters.[42]

The decision reaches considerably further than the 2007 Statement of the London Convention, declaring that "[s]uch studies should only be authorized if justified by the need to gather specific scientific data, and should also be subject to a thorough prior assessment of the potential impacts of the research studies on the marine environment, and be strictly controlled."[43]

Subsequently, in 2010, in one of the boldest political moves to date with respect to geoengineering governance, the subsidiary scientific body to the CBD agreed to forward to the full COP a proposed temporary ban on all forms of climate-related geoengineering on the basis of the precautionary principle. In relevant part, the proposal called upon the COP to:

> Ensure. . .on ocean fertilization and biodiversity and climate change, and in accordance with the precautionary approach, that no climate-related geo-engineering activities take place until there is an adequate scientific basis on which to justify such activities and appropriate consideration of the associated risks for the environment

---

[42] *Id.* § C(4).

[43] *Id.*

> and biodiversity and associated social, economic and cultural impacts.[44]

The CBD proposal represented the first time a UN body had directly called for the development of a geoengineering governance regime since the adoption of the ENMOD Treaty in 1977. The proposed ban was considered by the COP at its next full meeting, during which time the COP declined to adopt a direct ban, but voted to approve a decision *inviting* Parties and other non-Party governments to agree:

> That no climate-related geo-engineering activities . . . that may affect biodiversity take place, until there is an adequate scientific basis on which to justify such activities and appropriate consideration of the associated risks for the environment and biodiversity and associated social, economic and cultural impacts, with the exception of small scale scientific research studies that would be conducted in a controlled setting. . .and only if they are justified by the need to gather specific scientific data and are subject to a thorough prior assessment of the potential impacts on the environment.[45]

Despite the political significance of the invited ban, the substantive value remains unclear and untested. Left unanswered are questions as to what types of activities would be deemed to affect biodiversity, and at what point there will be an "adequate scientific basis" and "appropriate consideration of the associated risks" to justify moving forward with geoengineering activities.

In both the case of the London Convention and Protocol and of the CBD, the emphasis is on the need for precaution, given the depth of scientific uncertainty surrounding both the direct and indirect effects associated with geoengineering activities. In neither case do the institutions call for a permanent ban; rather, the dialogue in both forums emphasizes the need for further research and for more organized decision making and governance processes—something that continues largely to be lacking in international law.

While there continues to be an absence of directly governance mechanisms relevant to geoengineering, there are a number of principles of international law that could guide decision making in

---

[44] Convention on Biological Diversity, *Report of the Fourteenth Meeting of the Subsidiary Body on Scientific, Technical and Technological Advice*, § XIV/5(1)(8)(w), UNEP/CBD/COP/10/3, (June 20, 2010), http://www.cbd.int/doc/meetings/cop/cop-10/official/cop-10-03-en.pdf.

[45] *Id.* ¶ 8(w) (emphasis added). *See also* Convention on Biological Diversity, *New and Emerging Issues*, ¶ 4 (Oct. 29, 2010), UNEP/CBD/COP/10/L.26; Convention on Biological Diversity, *Marine and Coastal Biodiversity*, ¶ 13(e), 57–62.

the geoengineering context. The 1982 U.N. Convention on the Law of the Sea (UNCLOS)[46], for example, creates a general obligation to protect and preserve the marine environment and prohibits pollution of the marine environment. UNCLOS contains a number of provisions relating to protection of the marine environment, including obligations for states concerning use of marine natural resources. UNCLOS Article 194, for example, obligates states to take all measures necessary to prevent, reduce, and control pollution from any source and, thus, may be applicable to many geoengineering techniques (e.g., ocean fertilization). Similarly, UNCLOS Article 196 requires states to control pollution from the use of technologies under their jurisdiction and control. On the other hand, UNCLOS specifically allows scientific research on the high seas. While UNCLOS is "not universally applicable", many of the key provisions relating to protection of the marine environment "are generally considered to be part of customary international law and, consequently, binding on all states."[47] Thus, while neither being directive, nor creating a comprehensive regime for geoengineering, UNCLOS provides principles that may inform both discrete decision making processes, as well as the development of a more comprehensive set of rules or principles applicable to geoengineering.

Beyond UNCLOS, other principles (and emerging principles) of international law, including the precautionary principle, the principle of prevention of harm, the prohibition on transboundary harm, as well as general obligations under international law with respect to duties to cooperate and undertake environmental impact assessments provide background principles against which geoengineering projects should be assessed, and within which emerging geoengineering governance systems should be contextualized.

B. Emerging Public and Private Efforts to Govern Geoengineering

International legal bodies are not the only ones that have begun to coalesce around the idea of creating a set of guiding rules or norms for geoengineering. Around the time that the parties to the London Convention and the CBD were developing statements on geoengineering, a group of independent researchers were meeting in California to discuss the development of a "voluntary code of conduct" for geoengineering. This meeting was sponsored by the Climate Response Fund, a group formed in 2009 to "foster discussion of climate intervention research (sometimes called geoengineering or

[46] U.N. Convention on the Law of the Sea arts. 192, 193, Dec. 10, 1982, 1833 U.N.T.S. 3 [hereinafter 1982 UNCLOS].

[47] Scott, *supra* note 6, at 333–45.

climate engineering) and to decrease the risk that these techniques might be called on or deployed before they are adequately understood and regulated."[48] The Conference, entitled "The Asilomar International Conference on Climate Intervention Technologies," was coordinated with the intent of allowing the participants—there by invitation only—to work jointly to "develop norms and guidelines for controlled experimentation on climate engineering or intervention techniques."[49]

The Conference modeled itself after the 1975 Asilomar Conference on Recombinant DNA, wherein the participants agreed upon a series of restrictions and conditions that would guide their research. The original Asilomar Conference in 1975 was widely heralded as a positive example of self-regulation and was used as a model for the 2010 Asilomar International Conference on Climate Intervention Technologies.

The 2010 Conference stated its goal as "minimiz[ing] the risks associated with scientific research on climate intervention or climate geoengineering, much as the 1975 Asilomar Conference on Recombinant DNA successfully modeled safe and appropriate laboratory management methodologies." The Asilomar Conference focused exclusively on the development of risk reduction guidelines for climate intervention experiments. The creation of the Climate Response Fund and the organization of the Asilomar Conference likely were prompted by the rapid onset of attention focused on geoengineering in the preceding years.

The Asilomar Conference was preceded by a flurry of activity on the part of key scientific organizations, including the American Meteorological Society (AMS) and the American Geophysical Union (AGU), both of which called for research into all aspects of geoengineering, as well as a decision by the Royal Society of the United Kingdom—a leading global scientific body—to launch a major study into the governance of geoengineering.

The AMS and AGU statements helped usher in a new phase of geoengineering research, wherein research in this area became less taboo and more mainstream as geoengineering efforts were described as being able to "contribute to a comprehensive risk management strategy to slow climate change and alleviate some of its negative impacts." To this, both professional societies called for "adequate research, appropriate regulation, and transparent deliberation" on the matter and recommended:

---

[48] THE CLIMATE RESPONSE FUND, http://www.climateresponsefund.org/ (last visited Apr. 1, 2011).

[49] *Id.*

1. Enhanced research on the scientific and technological potential for geoengineering the climate system, including research on intended and unintended environmental responses.
2. Coordinated study of historical, ethical, legal, and social implications of geoengineering that integrates international, interdisciplinary, and intergenerational issues and perspectives and includes lessons from past efforts to modify weather and climate.
3. Development and analysis of policy options to promote transparency and international cooperation in exploring geoengineering options along with restrictions on reckless efforts to manipulate the climate system.[50]

Around the same time the AMS and AGU released their policy statements encouraging geoengineering research, Professor John Shepherd, the chair of the UK Royal Society's *Geoengineering the Climate* report,[51] declared the need to "consider beforehand what legislative mechanisms and guidelines are needed, to ensure that any [geoengineering] research that is undertaken will be done in a highly responsible and controlled manner with full international agreement where necessary."[52] To this end, the Royal Society initiated a study into the governance of geoengineering with the objective of agreeing upon a series of recommendations and best practices for the governance of research and deployment of SRM techniques, cautioning of the possible need for international agreement around the governance of these technologies.

The Royal Society's new study was launched in response to the release, just ten days before, of a report on *The Regulation of Geoengineering*, which was published by the United Kingdom's House of Commons. In this report, the Science and Technology Committee to the House of Commons declared the need to develop a common regulatory regime for geoengineering. In calling for the "groundwork for regulatory arrangements to begin" the Committee determined that:

---

[50] *Geoengineering the Climate System: A Policy Statement of the American Meteorological Society*, AM. METEOROLOGICAL SOC'Y (July 20, 2009), https://www.ametsoc.org/index.cfm/ams/about-ams/ams-statements/archive-statements-of-the-ams/geoengineering-the-climate-system-2009/.

[51] THE ROYAL SOC'Y, GEOENGINEERING THE CLIMATE: SCIENCE, GOVERNANCE AND UNCERTAINTY (2009), https://royalsociety.org/~/media/royal_society_content/policy/publications/2009/8693.pdf.

[52] THE ROYAL SOC'Y, POLICY PROJECTS: SRM GOVERNANCE INITIATIVE, https://royalsociety.org/topics-policy/projects/solar-radiation-governance/.

> Geoengineering techniques should be graded with consideration to factors such as trans-boundary effect, the dispersal of potentially hazardous materials in the environment and the direct effect on ecosystems. The regulatory regimes for geoengineering should then be tailored accordingly. The controls should be based on a set of principles that command widespread agreement—for example, the disclosure of geoengineering research and open publication of results and the development of governance arrangements before the deployment of geoengineering techniques.[53]

The House of Commons report offered a careful look at the risks and possibilities of geoengineering. It concluded that there is a regulatory gap around geoengineering and that this gap should be filled through the creation of a new international framework for the regulation of geoengineering in order to ensure "legitimacy; scientific standards; oversight mechanisms; and management of environmental and trans-boundary risks."[54]

Parallel to the House of Commons' consideration of the governance challenges associated with geoengineering, the U.S. House of Representatives' Committee on Science and Technology undertook an eighteen-month inquiry, including three public hearings, into the issue of climate engineering. The results of this inquiry were published five months after the release of the House of Commons report and within days of the CBD's decision to temporarily ban geoengineering. In contrast to the CBD's call for a temporary moratorium on geoengineering, the Committee's 2010 report[55] called for a more comprehensive and formal research agenda. While not advocating the use of geoengineering, the House Report called for an active and engaged debate on the science and policy of climate engineering. In releasing the report, Committee Chairman Bart Gordon stated that the report was "in no way meant as an endorsement of climate engineering" before declaring that:

> Climate engineering carries with it a tremendous range of uncertainties and possibilities, ethical and political concerns, and the potential for catastrophic side effects. I want to be absolutely clear that I am not in favor of

---

[53] SCIENCE & TECHNOLOGY COMMITTEE, THE REGULATION OF GEOENGINEERING: FIFTH REPORT OF SESSION 2009–10, 2010, H.C. 221, at 3 (U.K.) [hereinafter HOUSE OF COMMONS REPORT].

[54] *Id.* at 29.

[55] CHAIRMAN BART GORDON, COMM. ON SCI. & TECH., 111TH CONG., ENGINEERING THE CLIMATE: RESEARCH NEEDS AND STRATEGIES FOR INTERNATIONAL COLLABORATION (2d Sess. 2010), https://www.washingtonpost.com/wp-srv/nation/pdfs/Geongineeringreport.pdf (hereinafter House Report).

> deploying climate engineering. . . . However, if we find ourselves passing an environmental tipping point, we will need to have done research to understand our options. . . . We need healthy debate, a transparent process, clear action on emission reductions, and sound scientific research to provide a solid foundation for the tough decision-making that climate change will demand in the future.[56]

The House Report offered background information on geoengineering, which it refers to as climate engineering, including information about existing capacity within the U.S. federal agencies to inform geoengineering science as well as information about ongoing exploratory research worldwide. The report then explores key research needs and the tools needed to support this research at the federal level before offering models for how to organize future research. Finally, the House Report offered recommendations for how to proceed. Citing the need to understand the "most effective and risk-averse climate strategies" in advance of a potential "climate emergency," the Report declared that "it is the opinion of the Chair that broad consideration of comprehensive and multi-disciplinary climate engineering research at the federal level begin as soon as possible in order to ensure scientific preparedness for future climate events." The House Report also offered a series of recommendations. For example, the House Report recommended that:

- the global climate science and policy communities should work towards a consensus on what constitutes a 'climate emergency' warranting deployment of SRM technologies;
- there must ultimately be an international consensus on climate engineering terminology that will best communicate the strategies and desired effects to the scientific community, policy makers and the public;
- any federal climate engineering research program should leverage existing facilities, instruments, skills and partnerships within federal agencies;
- governments should make public engagement a priority of any climate engineering effort;
- further collaborative work between national legislatures on topics with international reach, such as climate engineering, should be pursued; and

---

[56] Press Release, Comm. on Sci. & Tech., *Chairman Gordon Releases Report on Climate Engineering* (Oct. 29, 2010), https://justfacts.votesmart.org/public-statement/570038/chairman-gordon-releases-report-on-climate-engineering.

- the U.S. Government should press for an international database of climate engineering research to encourage and facilitate transparency and open publication of results.

These recommendations support the ultimate conclusion that "policymakers should begin consideration of climate engineering research now to better understand which technologies or methods, if any, represent viable stopgap strategies for managing our changing climate and which pose unacceptable risks."[57] In advocating an open and coordinated geoengineering research agenda, the Chair cautioned against the adoption of a moratorium on research suggesting that a ban on research would conflict with principles of scientific freedom and impede accountability.

Unlike the U.K. House of Commons report, which focused on governance, the U.S. House of Representatives report focused primarily on research, leaving the question of the governance largely unaddressed. The report did suggest, however, both that there is a need for research and regulation to develop in parallel and that certain climate engineering research will require regulation and government oversight. The specifics of which types of climate engineering projects will need regulation, and what form of regulation and oversight will be needed are left open.

With a few exceptions, described below, the federal government has remained relatively agnostic and silent on the issue of geoengineering governance. The National Academy of Science, however, in 2015, in one of the most comprehensive reports to date on climate intervention highlighted the continuing governance gap and called for a more open and engaged conversation around the topic of governance, declaring that "'[g]overnance' is not a synonym for 'regulation'" and emphasizing that:

> [g]iven the perceived and real risks associated with some types of albedo modification research, open conversations about the governance of such research, beyond the more general research governance requirements, could encourage civil society engagement in the process of deciding the appropriateness of any research efforts undertaken.[58]

Then, in 2017, Congressman Jerry McNerney (CA) called for a hearing in the House Committee on Science, Space & Technology to gather further information on geoengineering. Subsequently, Representative McNerney introduced a bill, the Geoengineering

---

[57] *Id.* at ii.

[58] National Academy of Sciences, *supra* note 31, at 12.

Research Evaluation Act. The bill called for the National Academies to study and report on a geoengineering research agenda to advance understanding of albedo modification strategies and to provide guidance on governance mechanisms for the proposed albedo modification research agenda.[59] The bill never received a vote. However, in 2021, the National Academies released a new report providing a series of detailed recommendations on solar geoengineering research and research governance.[60] With respect to solar geoengineering, the report recommends that:

> the U.S. federal government should establish—in coordination with other countries transdisciplinary, solar geoengineering research program. This program should be a minor part of the overall U.S. research program related to responding to climate change. The program should focus on developing policy-relevant knowledge, rather than advancing a path for deployment, and should be subject to robust governance. . . The program should, from the outset, prioritize development of international coordination and co-development of research with other countries.[61]

Thus, despite increasing interest, geoengineering research and deployment remains largely unregulated and, even, ungoverned at the international and domestic levels. An increasing number of public and private actors are attempting to shape the emerging field. These actors approach geoengineering from different perspectives and with different end-goals. Participants in the Asilomar Conference, for example, seek a system of self-regulation. In contrast, the U.K. House of Commons recommends a top-down, international framework for regulation. The 2010 report by the U.S. House, meanwhile, advocated a formal and organized climate engineering research agenda, but avoided extensive discussions about associated regulatory needs. At the international level, the Parties to the CBD proposed de facto and outright moratoriums on certain geoengineering practices, but primarily as a result of existing

---

[59] H.R. 4586, 115th Cong. (1st Session) (Dec. 7, 2017).

[60] National Academies of Sciences, Engineering, and Medicine 2021. *Reflecting Sunlight: Recommendations for Solar Geoengineering Research and Research Governance*, The National Academies Press (2021). https://doi.org/10.17226/25762. This followed a two-volume report issued by the National Academies in 2015 outlining the research needs and ethical questions surrounding climate intervention, including CCS and albedo modification.

[61] *Id.* at 8. The report further recommends that the program "establish robust mechanisms for inputs from civil society and other key stakeholders", "be regularly reviewed and assessed by a diverse, inclusive panel of experts and stakeholders (including consultation with international counterparts) to determine whether continued research is justified and, if so, how goals and priorities should be updated", and include " 'Exit ramps' " . . . to terminate a research activity, for example, if it is deemed to pose unacceptable physical, social, geopolitical, or environmental risks." *Id.* at 9.

knowledge gaps. While there continue to be vocal critics of any and all geoengineering research, a consensus appears to be emerging that geoengineering research is an integral part of a larger climate research agenda, but that the question of which geoengineering technologies are effective, permissible, desirable, and allowable remains largely unanswered and unanswerable absent more research. Equally, the question of what type of governance regime is needed at the international level does not yet appear to be ripe, with no consensus yet emerging around this question.

C. Federal Regulation of Geoengineering in the United States

The United States does not have a comprehensive legislative or regulatory regime for geoengineering research or deployment. In the most comprehensive government report to date, the House of Representatives report avoided any in-depth discussion of what a future geoengineering governance regime could or should look like. Yet, as geoengineering research and proposals become more sophisticated and more likely to be deployed, future regulatory controls will be needed—possibly needing to be developed rapidly. The 2010 House report hinted at this conundrum, cautioning against overly restrictive regulations while simultaneously warning that the EPA, as the federal agency with primary authority over environmental concerns, must keep up with the evolving science and be ready to regulate should the need arise.

In the interim, however, certain geoengineering techniques are already the subject of evolving regulatory controls. This is particularly true with respect to carbon capture storage and sequestration efforts.

The primary way that the United States supports carbon capture storage and sequestration efforts is through a tax credit.[62] As of 2021, this tax credit provides $36 per ton of CO_2 stored in geological reservoirs or $24 per ton of CO_2 used in enhanced oil recovery or other purposes.[63] The credit applies to power plants that capture over 500,000 tons of CO_2 per year and to industrial facilities that capture over 100,000 tons of CO_2 per year.[64]

---

[62] IRC, *Credit for Carbon Oxide Sequestration*, 26 U.S.C. § 45Q.

[63] *See* Moch et al., *supra* note 24.

[64] *Id.* (further noting that the tax credit: "is set to increase each year until 2026, when it will provide facilities that began construction prior to 2026 with $50 per ton of CO_2 stored in geologic reservoirs and $35 per ton of CO_2 used for EOR or other processes" and noting that "[r]ecently introduced bipartisan legislation, the CCUS Tax Credit Amendments Act of 2021, proposes to increase the 2026 value of the 45Q tax credit from $50 to $120 per ton CO_2 for facilities that sequester CO_2 and from $35 to $75 per ton CO_2 for EOR facilities.")

In addition, the federal government is engaged in active research and development of clean coal technologies, including the use of CCS. For example, in February 2010, President Obama created the *Interagency Task Force on Carbon Capture and Storage* and tasked the group with developing a comprehensive and coordinated Federal strategy to speed the commercial development and deployment of clean coal technologies. The primary focus of the Task Force was on facilitating the widespread, economically efficient deployment of CCS. The Task Force concluded its work by providing a series of recommendations to the President on how to overcome barriers to the widespread, cost-effective deployment of carbon capture and storage that have informed subsequent federal efforts to support CDR research and deployment.

Anticipating the growth of CCS projects and associated sequestration needs, using its authority under the Safe Drinking Water Act (SDWA), the EPA developed a set of rules related to carbon sequestration in underground geological formations. The new rules, finalized in December 2010, were developed under the SDWAs Underground Injection Control (UIC) Program and include the creation of a new class of wells, Class VI wells. Class VI wells are designed to protect underground sources of drinking water and ensure the long-term geologic sequestration and storage of CO_2. The Program is also designed to ensure that these wells are appropriately sited, constructed, tested, monitored, funded, and closed.

In addition to the new rules under the SDWA, the EPA revised the Resource Conservation and Recovery Act (RCRA) to exclude carbon dioxide injection streams from the definition of hazardous waste. RCRA regulates the generation, transmission, storage, and disposal of solid wastes. The question before the EPA under RCRA was how streams of CO_2 injected for geologic sequestration should be classified under existing solid and hazardous waste requirements. CO_2 has not historically been classified as a hazardous waste under RCRA. In its final rule, issued on January 3, 2014, the EPA conditionally excluded CO_2 streams that are hazardous from the definition of hazardous waste, provided that the CO_2 streams are captured from emission sources, are injected underground via Class VI SDWA wells, and meet certain other conditions. EPA based this decision on the belief that so long as these CO_2 streams are safely managed, they do not pose a substantial risk to human health or the environment, and thus do not need to be subject to additional regulatory controls.

A related piece of environmental legislation that might apply to the release of CO_2 streams produced through CCS projects is the Comprehensive Environmental Response, Compensation, and Liability Act (CERCLA). CERCLA, more commonly known as the

Superfund, provides for the cleanup of contaminated sites where there have been releases of hazardous substances. CERCLA creates a very wide liability net and allows federal, state, and local governments and private landowners to bring liability claims for the recovery of costs associated with the cleanup of contaminated sites. While CO_2 is not a listed hazardous substance, the accidental release of which will trigger CERCLA cleanup and liability provisions, it is possible that a CO_2 stream produced during the CCS process may contain a listed hazardous waste or could interact with other substances in such a way as to produce a recognized hazardous waste, thus resulting in CERCLA liability. However, there is a specific CERCLA liability exemption for "federally permitted releases", and a Class VI permit may provide such an exception if the CO_2 stream is released in accordance with the permit requirements, making CERCLA liability unlikely under the existing legal regime.

Another legal tool that is potentially applicable to both CCS and other types of geoengineering projects is the National Environmental Policy Act (NEPA). NEPA is a law guiding decision making processes that involve major federal actions that could affect the environment. Pursuant to NEPA, any federal agency undertaking a "major Federal action significantly affecting the quality of the human environment" must complete an Environmental Impact Statement (EIS). Completing an EIS is a rigorous process that requires significant research and analysis of proposed actions and their consequences. Among other things, the EIS must evaluate the proposed project, alternatives to the projects, and any reasonably foreseeable direct and indirect environmental effects of the proposed action. NEPA is applicable in the context of CCS because many of these projects, and resulting pipeline projects, occur on federal and Tribal lands or require federal agency approval or financial assistance. While EISs are largely informational tools and do not dictate what actions can and cannot take place, they are important tools for purposes of public participation, transparency, information sharing, and decision making.

In addition to the regulatory work that the EPA and other federal agencies are undertaking under the SDWA, RCRA, and NEPA, a number of federal agencies, including the Department of Energy Fossil Energy Office and the National Energy Technology Laboratory, are involved in a wide array of CCS research and development programs nation-wide. The U.S. Department of Transportation is in charge of regulating CO_2 pipeline operations under the Interstate Commerce Act and Hazardous Liquid Pipeline Act, and the Federal Energy Regulatory Commission (FERC) and the Surface Transportation Board regulate applicable pipeline tariff rates and access.

These efforts have been bolstered by executive and administrative actions taken by the Biden Administration. The Biden Administration has set a goal of reaching net zero economy-wide greenhouse gas emissions by 2050.[65] Carbon dioxide removal is seen as being an essential tool to achieve this ambitious goal. To this end, in March 2021, the DOE approved $24 million in funding for direct air capture research. That same year, in October, the U.S. Department of Energy (DOE) announced that it was allocating $45 million in funding for 12 projects to advance point-source carbon capture and storage technologies that will be able capture at least 95% of carbon dioxide emissions generated from natural gas power and industrial cement and steel facilities[66]. In May 2022, Congress helped enhance these efforts by allocating almost $12 million towards carbon capture and storage programs under the Build Back Better Act, while the DOE, announced another $2.3 billion in funding to advance CDR projects.[67]

The emerging federal regime, as minimal as it currently is, should be viewed in conjunction with sub-federal efforts to regulate a variety of aspects of CCS. In the United States, much of the existing body of oil and gas law—including production, storage, and transportation phases—is state-based. Thus, it is no surprise that a number of states have begun to issue rules and judicial decision related to the use of the surface and subsurface for CO_2 pipelines and sequestration. For example, more than twenty states have enacted legislation that relates to CCS. The purpose of the state legislation varies significantly, but many of the state policies focus on creating incentives for CCS projects and establishing basic permitting and property rules for CCS sites.

Because the federal regulatory regime applicable to CCS is relatively thin, the states continue to have ample room within which to structure incentive and property-based regimes. Moving forward, the balance between federal and state law is likely to remain relatively stable absent significant efforts on the part of the EPA to

---

[65] White House, *Fact Sheet: President Biden Sets 2030 Greenhouse Gas Pollution Reduction Target Aimed at Creating Good-Paying Union Jobs and Securing U.S. Leadership on Clean Energy Technologies* (April 2021). https://www.whitehouse.gov/briefing-room/statements-releases/2021/04/22/fact-sheet-president-biden-sets-2030-greenhouse-gas-pollution-reduction-target-aimed-at-creating-good-paying-union-jobs-and-securing-u-s-leadership-on-clean-energy-technologies/.

[66] U.S. Department of Energy, *DOE Invests $45 Million to Decarbonize the Natural Gas Power and Industrial Sectors Using Carbon Capture and Storage* (Oct. 6, 2021), https://www.energy.gov/articles/doe-invests-45-million-decarbonize-natural-gas-power-and-industrial-sectors-using-carbon.

[67] U.S. Department of Energy, *Biden-Harris Administration Announces Over $2.3 Billion Investment to Cut U.S. Carbon Pollution* (May 2022), https://www.energy.gov/articles/biden-harris-administration-announces-over-23-billion-investment-cut-us-carbon-pollution.

revise the existing regulatory regime for site development and management, underground wells, and hazardous wastes associated with CCS projects.

Moreover, while SRM research and governance efforts continue to lag behind CDR, in 2020, Congress appropriated $4 million for NOAA's Office of Oceanic and Atmospheric Research (OAR) to investigate "Earth's radiation budget" and "solar climate interventions". This budget line was increased to $9 million in 2021. And in 2022, Congress directed NOAA to allocate at least $9 million in funding for continued study of stratospheric conditions and the Earth's radiation budget and to study "the impact of the introduction of material into the stratosphere from changes in natural systems, increased air and space traffic, and the assessment of solar climate interventions."[68] Congress further mandated that NOAA develop an interagency program to "manage near-term climate hazard risk and coordinate research in climate intervention" and directed the agency to produce a five-year plan for "scientific assessment of solar and other rapid climate interventions in the context of near-term climate risks and hazards."[69]

With respect to other types of geoengineering, the legal landscape remains relatively wide-open.

III. Conclusion

Humans have been interfering with the climate system for centuries. Many of the proposed geoengineering techniques, however, differ in intent, scale, and impact than past practices. This raises the questions: how do we regulate this emerging field and should we? Arguably, geoengineering is too important to be left stranded in a legal void, but as this chapter has explored, there is little consensus about how to govern the emerging field absent improved understanding of the different proposed techniques. In addition, the wide variety of activities that fall under the umbrella term of geoengineering—including everything from CCS, afforestation, and painting roof tops white to injecting aerosols into the stratosphere—makes it difficult to conceptualize how a top-down framework could cover the wide variety of activities, and attendant risks and benefit, that could take place under this heading.

---

[68] H.R. 2471, 117th Cong. (2nd Sess. 2022), https://www.govinfo.gov/content/pkg/CPRT-117HPRT47047/html/CPRT-117HPRT47047.htm.

[69] *Id.* ("The report shall include: (1) the definition of goals in relevant areas of scientific research; (2) capabilities required to model, analyze, observe, and monitor atmospheric composition; (3) climate impacts and the Earth's radiation budget; and (4) the coordination of Federal research and investments to deliver this assessment to manage near-term climate risk and research in climate intervention.")

The range of potential technologies varies so significantly that it may be illogical to attempt to create uniform regulatory systems for geoengineering at either the domestic or international level. Instead, perhaps, the focus could be on developing a set of normative principles that can guide decision making processes and allow for the development of technique-specific rules, such as with CCS. Candidate principles could include, for example, transparency and public participation, pollution prevention, duties of notification, independent monitoring and assessment, and intergenerational equity.

Regardless of how the governance system evolves, in thinking about geoengineering within the larger context of climate change law, it will be important to consider how to contextualize geoengineering efforts within the framework of ongoing efforts to mitigate and adapt to climate change. How do geoengineering efforts fit in? What are our comparative perceptions of risk and responsibility with respect to geoengineering, mitigation, and adaptation? For example, how can we compare the benefits and costs of a stratospheric aerosol proposal with a proposal to rapidly expand nuclear energy capacity so as to reduce greenhouse gas emissions?

With climate change, the global community is hemmed in by existing systems of law and is entrenched in a decades-long debate. With geoengineering, the governance slate remains relatively clean. The evolution of domestic and international governance geoengineering regimes is likely to become an increasingly active and important area of legal development in the decades to come.

Further Readings

Kelsi Bracmort & Richard K. Lattanzio, Cong. Research Serv., R41371, *Geoengineering: Governance and Technology Policy* (2013).

William C.G. Burns, *Climate Geoengineering: Solar Radiation Management and its Implications for Intergenerational Equity*, 4 STAN. J.L. SCI. & POL'Y 37 (2011).

Cinnamon Carlarne, *Arctic Dreams and Geoengineering Wishes: The Collateral Damage of Climate Change*, 49 COLUMBIA JOURNAL OF TRANSNATIONAL LAW 602 (2011).

Congressional Research Service, *Carbon Capture and Sequestration (CCS) in the United States* (Oct. 18, 2021), https://crsreports.congress.gov.

Paul J. Crutzen, *Albedo Enhancement by Stratospheric Sulfur Injections: A Contribution To Resolve a Policy Dilemma?*, 77 CLIMATIC CHANGE 211, 216 (2006).

Gary Ellem, *Carbon Capture and Storage is Unlikely to Save Coal in the Long Run*, THE CONVERSATION, https://theconversation.com/carbon-capture-and-storage-is-unlikely-to-save-coal-in-the-long-run-54182 (July 15, 2016).

Clare Heyward, *Time to Stop Talking about Climate Engineering*, https://ceassessment.org/time-to-stop-talking-about-climate-engineering-clare-heyward/ (Oct. 12, 2015).

David W. Keith, *Geoengineering the Climate: History and Prospect*, 25 ANN. REV. ENERGY & ENV'T 245, 261 (2000).

Jonas J. Monast, Brooks R. Pearson, Lincoln F. Pratson, *A Cooperative Federalism Framework Approach for CCS Regulation*, 7 ENVT'L & ENERGY L. & POL'Y J. 1, 4 (2012).

Steve Rayner et al., *The Oxford Principles* (2013), Climate Geoengineering Governance Working Paper Series: No. 1., https://www.homepages.ed.ac.uk/shs/Climatechange/Geo-politics/Oxford%20principles.pdf.

Alan Robock, *Stratospheric Aerosol Geoengineering*, 38 ISSUES IN ENVT'L SCIENCE AND TECH. 162 (2014).

Alan Robock, Allison Marquardt, and Ben Kravitz et al., *Benefits, Risks, and Costs of Stratospheric Geoengineering*, GEOPHYSICAL RESEARCH LETTERS, vol. 36 (October 2, 2009).

Stephen H. Schneider, *Abrupt Non-linear Climate Change, Irreversibility and Surprise*, 14 GLOBAL ENVTL. CHANGE 245 (2004).

Karen N. Scott, *International Law in the Anthropocene: Responding to the Geoengineering Challenge*, 34 MICH. J. INT'L L. 309, 322 (2013).

Elise Stull, Xiaopu Sun & Durwood Zaelke, *Enhancing Urban Albedo to Fight Climate Change and Save Energy*, 11 SUSTAINABLE DEVELOPMENT LAW & POLICY 1, Article 5 (2010).

David Takacs, *Environmental Democracy and Forest Carbon* (REDD), 44 ENVTL. L. 71 (2014).

David G. Victor, *On the Regulation of Geoengineering*, 24 OXFORD REV. ECON. POL'Y 322, 323 (2008).

TABLE OF CASES

INDEX

References are to Pages